AF622555

Sludge Thermal Hydrolysis

Application and Potential

Sludge Thermal Hydrolysis

Application and Potential

William Barber

IWA PUBLISHING

Published by **IWA Publishing**
Alliance House
12 Caxton Street
London SW1H 0QS, UK
Telephone: +44 (0)20 7654 5500
Fax: +44 (0)20 7654 5555
Email: publications@iwap.co.uk
Web: www.iwapublishing.com

First published 2020

British Library Cataloguing in Publication Data
A CIP catalogue record for this book is available from the British Library

ISBN: 9781789060270 (paperback)
ISBN: 9781789060287 (eBook)
ISBN: 9781789060294 (ePub)

Contents

Chapter 8

About the Author

William (Bill) Barber is a chartered engineer with a PhD and degree in Chemical Engineering. His interest in sewage treatment began with his PhD at Imperial College, London, under the supervision of Prof. David Stuckey where he investigated the fate of nitrogen and sulphur compounds in high rate anaerobic digestion. Bill extended this work at post-doctoral level to look at the treatment of low strength industrial wastewaters at low hydraulic retention times. Subsequently, Bill has worked in a variety of roles including consulting for multi-national companies, technical development and sales for various technologies, and for the UK's largest Water Company United Utilities where, as a technical advisor, he was part of the team that developed the award-winning thermal hydrolysis facility at Davyhulme, the largest plant in Europe. He has worked on projects in the UK, Europe, Asia, Australia, New Zealand, and North America. With respect to thermal hydrolysis, in addition to the project at Davyhulme, Bill undertook a study for the Asian Development Bank advising on the implementation of the technology in China. Subsequently, Bill was part of a team working on behalf of DC Water at Blue Plains on projects related to their facility. Bill has been invited to many conferences around the world as a keynote speaker to present on sludge and biosolids management and also helps in developing biosolids content for events. He has written numerous peer-reviewed articles, including a review on thermal hydrolysis, and is himself reviewer for several academic journals.

Preface

Around the world, the production of sewage sludge is increasing. This is due to various factors including growing population, increasing urbanization and greater investments in wastewater infrastructure. In addition, the composition of sewage sludge is evolving because of stricter environmental regulation. This change is towards the production of activated or biological sludge produced during oxidation of wastewater to destroy nutrients, away from primary sludge produced during initial settlement of the incoming solids in wastewater influent. Unfortunately, the processing of activated sludge is notoriously difficult and has been widely lamented in the scientific literature within 15 years of the commercialization of the activated sludge process in 1914. The principle difference between activated and primary sludge is the high content of extracellular polymer material found in activated sludge. This material makes the sludge viscous and difficult to transport. It also retains water which drastically limits the dewaterability potential for the sludge. Additionally, this material makes activated sludge difficult to degrade during anaerobic digestion. As the use of anaerobic digestion of sludge has been growing in interest due to the production of renewable energy amongst other benefits, so has the interest in improving the anaerobic biodegradability of activated sludge seen as the rate-limiting step in the anaerobic digestion of sewage sludge. Decades of research have been dedicated to this pursuit investigating a wide variety of, sometimes exotic, approaches. These approaches are almost always applied to the sludge prior to anaerobic digestion. Of these, thermal hydrolysis, the application of heat

Author: William Barber
doi: 10.2166/9781789060287_xi

in the range of 130-200°C has gained much attention especially since the installation of the first full-scale commercially operating plant in Hias, Norway. The principle benefits of the technology can be linked in part to the destruction of the extracellular material. This significantly reduces viscosity which helps overcome transport problems. This allows digestion plants to be routinely operated between 2 and 3 times the loading rates of plants where thermal hydrolysis is absent as a pre-treatment. Along with the solubilization of material in the sludge, thermal hydrolysis improves the biodegradability of activated sludge thereby increasing production of renewable energy in the form of biogas at the expense of digested sludge exiting the process. The destruction of extracellular material makes the sludge more compressible which improves the dewaterability of the digested sludge, ironically the original use of sludge thermal hydrolysis. The aim of this book is to help the reader gain an understanding of the thermal hydrolysis process with respect to sludge processing, and appreciate the benefits and challenges of using the technology.

Acknowledgements

I would like to thank Phil Pedros for checking through the text and calculations in this book and for providing useful comments and suggestions on the manuscript. In addition, I thank the staff at Cambi, especially Di Deng who provided useful insight on the thermal hydrolysis market in China. I extend my thanks to Sudhakar Viswanathan, of Kruger Veolia Water Technologies, who generously provided information on the facility at the Billund Biorefinery. Furthermore, I would like to thank Bill Brower and Chris Peot and the team I worked with at DC Water's Blue Plains facility. I am grateful for the support of Rick Lancaster, and the SBAP team with whom it was a privilege to work with during my time at United Utilities during the development of the facility at Davyhulme. Finally, I would like to thank my family and friends for their incredible and ongoing support, without which none of this would be possible.

Chapter 1

Introduction

The aim at present is a high gas yield and as rapid a digestion as is consistent with the production of an inoffensive product.

A. M. Buswell (1930)

1.1 INTRODUCTION

In his words above, Buswell succinctly summarizes the fundamental aim of sewage treatment. That of optimizing production of renewable energy in as small a facility as possible, without compromising on the generation of a high-value end product. Sixty-five years after these words were written, the first full-scale, commercially operating thermal hydrolysis facility was installed in Hias, Norway (Ødeby *et al.*, 1996). The facility is still in operation today with 98% average availability, 63% destruction of volatile matter and biosolids cake dewatered to 38% dry solids (DS) on a piston-press provided by Bucher Unipektin (Netteland *et al.*, 2015).

Today, thermal hydrolysis of municipal sludge is a mature technology and is considered mainstream by many practitioners. Many owners have gone on to buy multiple facilities and set their strategic plans for biosolids management around the technology. However, there have been many teething problems along the way, so much so that the concept of thermal hydrolysis was close to never becoming a reality. A great deal of experience has been learnt in the field, the importance of which should not be underestimated when considering a

Author: William Barber
doi: 10.2166/9781789060287_0001

technology supplier. Three factors have contributed to the development of not just thermal hydrolysis but pre-treatment technologies generally:

(1) The invention of the activated sludge process
(2) Lack of innovation in the design of municipal anaerobic digestion
(3) The long lifetime of anaerobic digestion infrastructure

Key to the success of pre-treatment is the activated sludge process to nitrify wastewater. In 1898, Scott-Moncrieff envisioned a future where wastewater treatment would comprise anaerobic digestion followed by nitrification under highly aerobic conditions (Scott-Moncrieff, 1898). However, attempts to treat wastewater by nitrification, by Dupre and Dibdin in London in the 1880s, and later by the aptly named Dr Drown in Lawrence, could not meet the required outcomes in a sufficient time period (Mohlman, 1917). However, by decanting sludge, rather than disposing of it, Ardern and his co-worker Lockett, managed to nitrify wastewater in a practical and reasonable time period by developing what they termed "activated sludge" (Ardern & Lockett, 1914). While Arden and Lockett answered key questions related to wastewater treatment in one of the 20th century's most influential inventions, they inadvertently created numerous new queries related to how to accommodate this new "activated" sludge. Zack and Edwards (1929) were lamenting the poor digestibility of this material. Even today, activated sludge (and variations of) remains the rate-limiting step of any sludge processing infrastructure, especially anaerobic digestion and dewatering. Activated (biological) sludge is fundamentally different to that produced during primary treatment. Unlike primary sludge, high quantities of extracellular polymers (which comprise carbohydrates, proteins, nucleic acids, lipids and humic acids, Goodwin & Forster, 1985; Urbain *et al.,* 1993) give activated sludge a gel–matrix structure which is highly viscous. During anaerobic digestion this results in mass-transfer limitations, making hydrolysis, rather than traditionally methanogenesis rate limiting. During dewatering, the water-retaining properties of the gel–matrix guarantee poor performance (Neyens *et al.,* 2004). Previous work looking at a variety of ways to destroy these extracellular materials concluded that only temperatures above autoclave (typically over 121°C) could destroy the problematic compounds (Barjenbruch & Kopplow, 2003). Ironically, the discovery of "activated" sludge was made at Davyhulme, which currently boasts Europe's largest thermal hydrolysis facility (described in Chapter 6).

Compared to activated sludge, primary sludge has greater quantities of lipids and fibres but far less protein and phosphorous. The following molecular formulae have been developed for typical primary and activated sludge with data from various sources to highlight the difference between typical primary and activated sludges:

- Primary sludge $C_{23}H_{35}O_8N$ (C:H:O:N = 61:8:28:3)
- Activated sludge $C_7H_{11}O_3N$ (C:H:O:N = 53:7:31:9)

From the composition data it is possible to evaluate the calorific value using a sludge-specific version of the Dulong equation (Technical Report, CEN/TR

13,767, 2004). The use of this equation gives calorific values of approximately 25,700 for primary and 21,800 kJ/kg volatile material for activated sludge. The greater carbon content (primarily from the lipids) results in primary sludge having approximately 15–20% more energy than sludge formed during aeration. The COD equivalence can be determined from:

For primary sludge

$$C_{23}H_{35}O_8N + 27O_2 \rightarrow 23CO_2 + 16H_2O + N$$

For activated sludge

$$C_7H_{11}O_3N + 7.5O_2 \rightarrow 7CO_2 + 4H_2O + N$$

Table 1.1 highlights the differences between primary and activated sludge calculated from the molecular formula above and from literature.

The following example highlights the limitation of activated sludge digestion.

Table 1.1 Differences between primary and activated sludge.

	Primary Sludge	Activated Sludge
COD/VS ratio	1.91	1.53
Oxidation state	−1.04	−1.43
Energy content (kJ/kg VS)	25,700	21,800
Gas yield (m^3/kg destroyed)	1.06	0.79
Gas composition (%CH_4:%CO_2)	63:37	68:32
Gas energy content (kWhr/kg destroyed)	7.147	5.713
Volatile solids destruction (%)	55–65	30–35
Dewaterability (% DS)	25 >	15–21
Protein (mg/g SS)	140–300	190–420
Polysaccharides (mg/g SS)	200	50–200
Lipids (mg/g SS)	100–650	30–100
Humics (mg/g SS)	80–140	60–200
Nitrogen (mg/g SS)	30	50
Phosphorous (mg/g SS)	15	60
Total EPS (mg/g SS)	75	130
of which protein (%)	44	58
of which polysaccharides (%)	8	10
of which humics (%)	48	32
EPS surface charge (meq/g SS)	0.18	0.33
Floc size (μm)	53	125

Based on theoretical calculations, also adapted in part from Zhen *et al.* (2017), Wang *et al.* (2019); and Metcalf & Eddy, *Wastewater Engineering, Treatment and Resource Recovery*, 5th Edition, 2014.

Question. *How much more energy does biogas contain from the anaerobic digestion of 1 kg VS primary sludge compared to an equivalent amount of activated sludge? Assume primary sludge is 55% biodegradable and activated sludge 30% degradable.*

Answer.
For primary sludge 1 kg VS is digested and 55% is converted to biogas.
Therefore, energy in the gas produced

$$= 1\,\text{kg VS} \times 55\%\ \text{destroyed} \times 7.147\,\text{kWhr/kg}_{\text{destroyed}}$$

$$= 3.931\,\text{kWhr}$$

For activated sludge 1 kg VS is digested and 30% is converted to biogas.
Therefore, energy in the gas produced

$$= 1\,\text{kg VS} \times 30\%\ \text{destroyed} \times 5.713\,\text{kWhr/kg}_{\text{destroyed}}$$

$$= 1.714\,\text{kWhr}$$

Consequently, biogas produced from anaerobic digestion of primary sludge contains over twice the energy content compared with the digestion of an equivalent quantity of activated sludge.

However, there are numerous variations of activated sludge treatment, and these yield numerous types of sludge. In addition, as wastewater legislation tightens further, the composition and quantity of activated sludge changes to that less amenable to anaerobic digestion. The length of time the sludge is aerated, (the sludge age), plays an important part in downstream processing of the sludge. Gossett and Belser (1982) studied the impacts of varying sludge age on the digestibility of activated sludge at 15 d hydraulic retention time. The researchers found that volatile solids destruction fell from 30% to 25% between 5 and 10 d sludge age showing that a shift from carbonaceous removal to nitrification would result in lower biogas generation from the activated sludge. Performance deteriorated with increasing sludge ages with approximately 20% destruction at 20 d sludge age and <15% at 30 d, these sludge ages being typical of extended aeration systems. Such systems are common in Australia and many are not preceded by primary treatment. Batstone and co-workers undertook a similar study to revisit the results in a modern Australian context (Batstone *et al.*, 2011). As with the previous work, this study found a direct correlation between anaerobic degradability of activated sludge and sludge age, and suggested that sludge ages in excess of 15 d would produce sludge which would most likely result in failure of a downstream digestion plant. In their study, digestibility of secondary sludge approximately halved between 5 and 20 d sludge age and fell by almost an additional third as aeration time was increased to 30 d.

The lack of innovation regarding municipal anaerobic digestion also provides fertile ground for pre-treatment technology. Looking through patents, it becomes clear that in the early 1900s it was well understood that staged anaerobic digestion performed better than equivalent parallel digestion. However, it was not until the 1970s that "acid" phase digestion was introduced. Even today, it is difficult to find designs based on staged digestion, which do not include using different temperature combinations, and many new build thermal hydrolysis facilities are combined with parallel digestion. Although most trophic groups found during anaerobic digestion are notoriously sluggish, negligible interest has been shown in the municipal sector to maintaining their presence within an anerobic digester. Recuperative thickening, where a fraction of the sludge – and therefore, microbial biomass – exiting the digester is returned is currently in vogue (Ostapczuk *et al.*, 2011; Yang *et al.*, 2017, 2015), but previously had been largely dormant since the original work of Torpey and Melbinger from (1967). Subsequently, the standard designs of anaerobic digestion plants are far from optimal and this provides abundant opportunities for processes which can easily improve performance within the limitations of the initial design.

1.1.1 Hydrolysis processes

The limitations related to the anaerobic digestion of activated-type sludges can be overcome somewhat by the use of "hydrolysis" processes installed upfront of the digestion plant. Hydrolysis with respect to anaerobic digestion, relates to the breakdown of complex organic polymers by extracellular enzymes of facultative or obligate anaerobic bacteria into their monomeric/oligomeric constituents small enough to allow their transport across the cell membrane. By doing this, complex material becomes more biodegradable and consequentially more biogas can be produced at a faster rate, shown by the biogas production profiles in Figure 1.1 which shows typical biogas response.

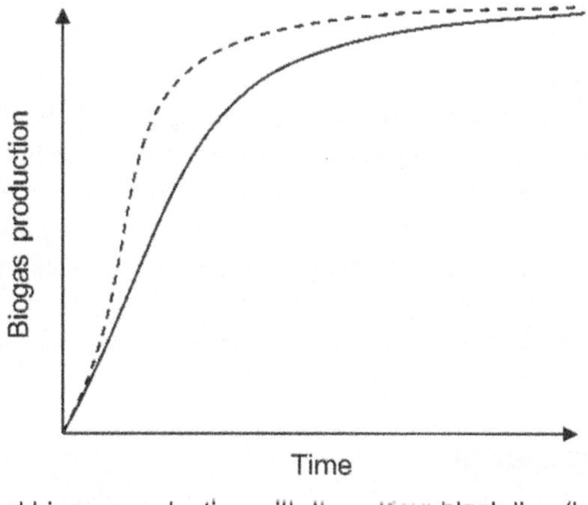

Figure 1.1 Typical biogas production with time. Key: black line (hydrolysis limited); dotted line (hydrolysis enhanced).

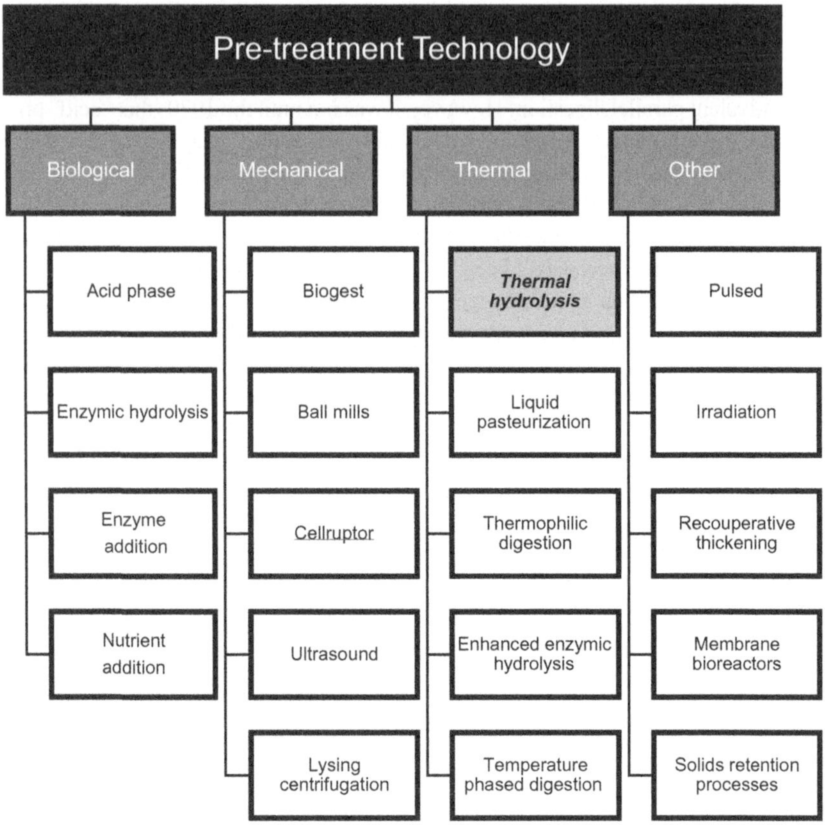

Figure 1.2 Hydrolysis technologies for sewage sludge.

The figure, representative of those found during determination of biodegradability of samples for digestion, shows that hydrolysis is effective to get more biogas especially at lower times, enabling digesters to be designed at lower retention times. With thermal hydrolysis, the retention time becomes limited by the capability of the digester to avoid cell washout, typically around 10 d.

These hydrolysis systems use different mechanisms to improve the rate of degradation of activated sludge, and fall loosely into four categories as shown in Figure 1.2.

The mechanism behind thermal hydrolysis is to use heat above autoclave temperature to improve hydrolysis.

1.1.2 Heating sludge

Thermal hydrolysis involves the application of heat to thickened or dewatered sludge produced during wastewater treatment at temperatures between 120°C and 200°C for a defined time period. The addition of this quantity of heat to sludge is not new, and early work was carried out on using heat to improve the dewaterability characteristics of sludge (Lumb, 1940, 1951). These systems had been developed to heat sludge to sub-critical conditions in the presence of an oxidant to temperatures between 200°C and 300°C at pressures between 10 and 200 bar. Commercially, these systems (sub-critical wet air oxidation (SCWAO)) have been around since the mid-20th century. The Zimpro process was developed in the 1930s and this was followed by the Porteus process in the 1960s. Like thermal hydrolysis, Zimpro used steam to heat sludge, but up to maximum operating temperatures between 150°C and 300°C. While operating at similar temperatures, the Porteus process used compressed air injected by an air compressor for heating with an energy release of approximately 0.85 kWhr/kg air added. In these systems, organics were oxidized, leaving acetic acid and ammonia in the liquid stream. The solids produced were dewatered to very high DS content typically prior to incineration. While these systems were beneficial with respect to destroying most of the organics in the water and being beneficial for dewatering, they suffered from issues surrounding odor, high operating costs and production of wastewater with refractory material within it. Subsequently, at the time of writing, only a handful of these types of facilities remain in operation globally. Incidentally, one of these plants, at Kenneth W. Hotz Water Reclamation Facility in Ohio, United States, has recently been decommissioned and replaced with thermal hydrolysis combined with anaerobic digestion.

Although most early interest was based on dewatering enhancements (Evert, 1972) the concept of applying heat to improve the biodegradability of sewage sludge was conceived later (Haug, 1977; Haug *et al.,* 1978; Stuckey & McCarty, 1984). Haug and co-workers (1978), discovered that the biodegradability of waste-activated sludge increased by 14% at 100°C, but by 60–70% at 175°C. However, it was noted that increasing temperature further decreased biodegradability due to the production of inhibitory materials, as did manipulating the hydrolyzed sludge with acids and alkalis at 175°C. Primary sludge did not appear to be overly influenced by THP but nor did it produce inhibitory compounds. The reported optimal operating conditions involve the application of temperature between 160°C and 180°C, for a time period of between 20 and 40 min (Bougrier *et al.,* 2006, 2008; Haug, 1977; Haug *et al.,* 1978; Li & Noike, 1992; Neyens & Baeyens, 2003; Stuckey & McCarty, 1984), although researchers have studied temperature ranges of 60–275°C, for 10–180 min.

Studying the pertinent literature (Bougrier *et al.,* 2008; Chu *et al.,* 2002; Lu *et al.,* 2014; Li & Yang, 2007; Lui *et al.,* 2012, as examples) it becomes apparent that as temperature increases, different reactions occur, these are summarized in Figure 1.3 (adapted from Barber, 2016).

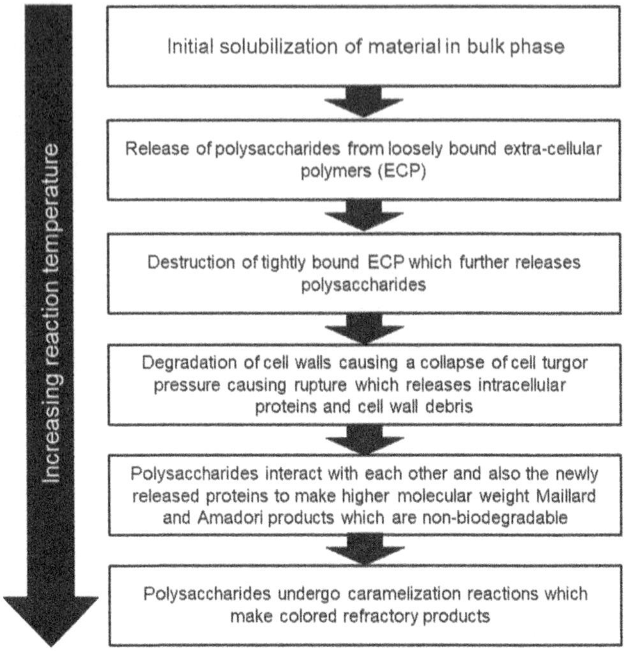

Figure 1.3 Sequence of reactions due to increasing temperature of sludge. (Adapted from Barber, 2016).

Increasing thermal hydrolysis reaction temperature up to optimum temperature range:

- Improves downstream sludge anaerobic digestibility
- Decreases apparent viscosity
- Increases solubility of carbohydrates
- Increases solubility of proteins
- Has negligible influence of solubility of lipids
- Reduces average particle size
- Increases potential for refractory compound formation (COD, nitrogen, color)

Increasing thermal hydrolysis reaction temperature beyond optimal temperature to sub-critical water range:

- Decreases downstream sludge anaerobic digestibility
- Significantly increases production of refractory material and color
- Further reduces viscosity
- Further improves dewaterability

The increasing production of refractory material (discussed in more detail in Chapter 3) with increasing temperatures is clearly evident in the work of Stuckey and McCarty (1984) who noted a continual deterioration in gas production from thermally hydrolyzed activated sludge which was 27% higher than a control at 175°C, similar at 250°C, and lower than the control above that temperature. At thermophilic conditions the impact of increasing temperature on reducing performance was further exacerbated. However, as well as increasing the quantity of refractory material, increasing the reaction temperature has also been linked to improved dewaterability (Evert, 1972; Higgins *et al.*, 2017) and enhanced biogas production (Lu *et al.*, 2014; Stuckey & McCarty, 1984), implying different optimal operating conditions exist depending on required project outcomes.

1.1.3 Principles of thermal hydrolysis

Traditionally, the process has been installed preceding anaerobic digestion, but more recently is being found downstream of the digestion process. The principle components of a thermal hydrolysis plant are shown in Figure 1.4.

Initially, sludge collected from wastewater treatment is thickened to anywhere between 10 and 30 percentage points DS to reduce the heating demands of the thermal hydrolysis plant itself (described in Chapter 2). Once in the thermal hydrolysis unit, sludge is heated to reaction temperature and concomitant pressure (typically 165°C at 7 bar) by the addition of live steam or heat exchangers, and is typically kept for a defined period, although some processes pass the sludge through in a continuous fashion. The heating of sludge causes solubilization of material within it, destroys materials such as extracellular polymers in activated sludge, and sterilizes the sludge from micro-organisms. Furthermore, it makes the sludge more compressible which makes dewatering of

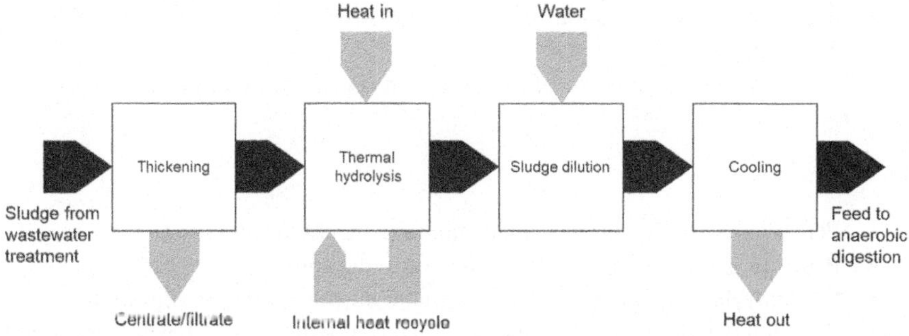

Figure 1.4 Typical components of a sludge thermal hydrolysis system.

the sludge easier. With batch systems, the sludge is then rapidly depressurized/flashed, or exploded to atmospheric pressure which causes destruction of the sludge particles. Particle destruction combined with solubilization make the sludge more amenable to subsequent processing resulting in increased performance during anaerobic digestion. The steam given off during the depressurization is generally recycled in batch-based systems to reduce overall heating requirements as will be discussed later. After processing, the sludge is removed from the plant at approximately 100°C and slightly lower dry solids due to dilution with the added steam. It is then diluted further using clean water or treated effluent to control the loading rate to the downstream anaerobic digestion plant. This addition also reduces the sludge temperature by approximately 20°C. Finally, the diluted hydrolyzed material is further cooled to the requirements of the digestion plant.

The main advantages and disadvantages of thermal hydrolysis are summarized, namely.

Advantages:

- Reduces sludge viscosity
- Significantly improves the biodegradability of waste-activated sludge
- Slightly improves the biodegradability of primary sludge
- Increases rate of biogas production
- Improves sludge dewaterability on all dewatering systems
- Sterilizes sludge providing pathogen-free biosolids
- Reduces odor and eliminates pathogen regrowth from dewatering
- Significantly reduces scum and foaming and produces conditions which do not encourage foaming
- Minimizes inhibition due to hydrogen sulfide
- Releases readily biodegradable COD which can be used for biological nutrient removal
- Significantly reduces downstream requirements for drying and other thermal processes
- Makes high concentration ammonia side-stream which is more suitable for advanced nutrient removal
- Reduces carbon footprint of treatment irrespective of end use for biosolids

Disadvantages

- Perceived as more complex than standard anaerobic digestion
- Parasitic energy demand with some configurations
- Greater release of ammonia and nutrients during anaerobic digestion
- Potential production of refractory material which may have detrimental impacts on wastewater standards

- May require operators with specialized training
- Sludge needs to be dewatered before hydrolysis to reduce energy demand
- Polymer dosage requirement may increase on a unit basis

Generally, the efficacy of thermal hydrolysis is inversely proportional to the initial biodegradability of the material (Wilson & Novak, 2009). Many researchers have investigated the influence on thermal hydrolysis on different types of sewage sludge and waste materials which are generally considered as substrates for co-digestion. With respect to sewage sludge, it is universally accepted that the biodegradability of sludge produced during activated sludge treatment or variants of it improves at a far higher rate than sludge collected during primary treatment.

1.1.4 Overview of influence on sewage sludge treatment

Table 1.2 summarizes the following influences of thermal hydrolysis from the literature (adapted from Barber, 2016):

These influences result in a cascade of further impacts, and their interactions are summarized in Figure 1.5.

The key impacts of thermal hydrolysis will be discussed in further detail in Chapter 3.

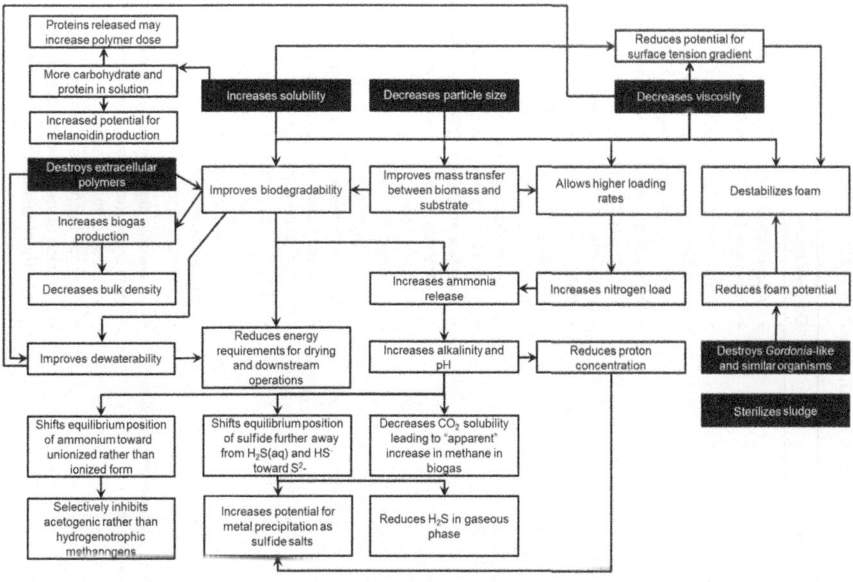

Figure 1.5 Cascading influences of thermal hydrolysis in sludge treatment.

Table 1.2 Influence of thermal hydrolysis.

Parameter	Observation	Reference
Alkalinity	• Approximately doubles from 780 to 1,580 mg/l at 170°C for 30 min	Wang *et al.* (2009)
	• Increase from 0.67 to 4.23 g/l at 165°C for 50 min	Han *et al.* (2017)
Anaerobic biodegradability	• Measured by gas production, increases with temperature between 130°C and 170°C (with constant reaction time of 30 min), with higher production of biogas noted after 10 d with temperature. However, little difference in biogas production noted in first 24 hr.	Lu *et al.* (2014)
	• Minor improvement in volatile solids reduction during digestion with increasing thermal hydrolysis processing temperatures. However, methane yield (m^3 methane/kg VS added) higher at 170°C compared to lower temperatures which were all similar.	Higgins *et al.* (2015)
	• WAS digestion significantly improved by 75–80% by over-pressurizing to 21 bar although no discussion as to the additional energy requirements of over-pressurization	Phothilangka *et al.* (2008)
	• Increase in gas production yield of over 100 l/kg VS fed at lab-scale with 20:80 primary: activated-sludge mixture.	Oosterhuis *et al.* (2014)
	• VSR of WAS increases from 26% to 42% (relative increase of 62%)	Van Dijk and de Man (2010)
	• Relative increase in WAS digestion of 55% at 140°C. When pre-treating sludge by thermal hydrolysis before digestion, Haug calculated a 25% increase in energy production compared to conventional digestion.	Haug *et al.* (1978)
	• Thermal pre-treatment at temperatures below 100°C revealed an increase of more than 30% in gas production at lower temperatures of 60 and 80°C, but the low-temperature pre-treatment necessitated a longer contact time than the high-temperature treatment.	Hiraoka *et al.* (1985)
	• Dropping thermal hydrolysis reaction temperature from 180°C to 160°C had no impact on digestion performance	Netteland *et al.* (2015)
	• Biogas production increases with temperature between 130°C and 190°C	Qiao *et al.* (2013)
	• Increasing reaction time from 30, to 60 and 90 min had no impact at temperatures between 130°C and 190°C, however, at 190°C increasing retention time resulted in a reduction in biogas production	Qiao *et al.* (2013)
	• Optimum biogas yield at 180°C for WAS concentrations of 1–7% by weight	Jeong *et al.* (2019)
Carbon footprint of biosolids management	• Anaerobic digestion reduces carbon footprint of biosolids processing. Adding thermal hydrolysis further reduces carbon footprint resulting in lowest carbon footprint regardless of end use of the biosolids	Barber (2008, 2017)
	• Thermal hydrolysis with digestion is beneficial with respect to reducing carbon footprint	Auty and Blake (2009)
	• Addition of thermal hydrolysis in Washington DC, along with deammonification for side-stream treatment resulted in a large reduction of carbon footprint	Willis *et al.* (2010)

	• When compared with numerous biosolids options, thermal hydrolysis combined with digestion and land recycling of the biosolids cake was the lowest carbon footprint	Asian Development Bank (2012)
Carbon source	• Can provide carbon source for nutrient removal immediately after hydrolysis with a C:N ratio of 12.5:1	Barlindhaug and Ødegaard (1996), Pinnekamp (1989)
Color	• Measured increase in melanoidin for kitchen waste and vegetable fruit waste	Liu *et al.* (2012)
	• Reducing thermal hydrolysis reaction temperature reduces color in digestate	Dwyer *et al.* (2008)
	• Increasing thermal hydrolysis reaction temperature increases color production in an approximately linear fashion from 2,700 mg PtCo/l at 140°C to 12,500 mg PtCo/l at 165°C	Dwyer *et al.* (2008)
	• Increasing adsorbance at 254 nm with increasing thermal hydrolysis reaction temperature	Wilson and Novak (2009)
	• Sludges with high fractions of activated sludge produced more color than those with higher fractions of primary sludge	Svennevik (2019)
Dewaterability	• Improves with increasing temperature between 130°C and 170°C in a linear fashion from approximately 27–32% DS	Higgins *et al.* (2015)
	• Increase from 25.2 to 32.7% DS at 180°C and 21 bar	Phothilangka *et al.* (2008)
	• Dewaterability improves with temperature between 180°C and 210°C	Evert (1972)
	• Improvement from 26% to 33.9%	Oosterhuis *et al.* (2014)
	• Centrate from thermally hydrolyzed digested sludge a third higher in COD and three times higher in ammonia compared to digested sludge	Barber (2010)
	• When thermally hydrolyzed digested sludge is mixed with sludge which has not been exposed to treatment the dewaterability of the mixture is slightly lower than predicted by weighed average of the separate sludges	Barber (2010)
	• Heat treatment of sludge improves dewaterability whether digested or not	Haug *et al.* (1978)
	• Dewaterability of thermally hydrolyzed raw sludge and post digested sludge far better than sludge which has been thermally hydrolyzed and then digested	Lumb (1951), Sherwood and Philips (1970)
	• Significantly improved by thermal treatment	Brooks (1968, 1970)
	• Reducing thermal hydrolysis operating temperature from 180°C to 160°C had no adverse impact on dewatering	Netteland *et al.* (2015)
	• Dewatering at higher temperature enhances dewatering potential	Svennevik (2019)
	• Dewatering improves with increasing temperature between 130°C and 190°C	Qiao *et al.* (2013)
	• Increasing reaction time from 30 to 60 and 90 min has no impact on dewaterability	Qiao *et al.* (2013)
Digestion temperature	• Pre-treatment better suited to meso- rather than thermophilic digestion at thermal hydrolysis temperature up to 120°C	Bi *et al.* (2013)
	• Although highest methane yield was at 53°C digestion	

(*Continued*)

Table 1.2 Influence of thermal hydrolysis (*Continued*).

Parameter	Observation	Reference
Dissolved organic nitrogen	• Increases with thermal hydrolysis reaction temperature for activated sludge from approximately 2,000 mg/l at 140°C to nearly 3,000 mg/l at 165°C	Dwyer *et al.* (2008)
	• Little influence for ammonia observed at full-scale	Tong *et al.* (2019)
	• Ammonia increases with reactor temperature, but total nitrogen remains uniform	Jeong *et al.* (2019)
	• No influence on TKN at 170°C for 30 min	Wang *et al.* (2009)
	• Increase in dissolved ammonium from 0.4 to 1.3 g/l at 160°C for 30 min	Chen *et al.* (2018)
	• Increase in dissolved protein from 0.2 to 15.05 g/l at 160°C for 30 min	Chen *et al.* (2018)
	• Approximately twice as high in primarily activated sludge compared to mixed sludge	
Extracellular polymers	• Completely destroyed by autoclaving (121°C)	Barjenbruch and Kopplow (2003)
	• Destroyed by thermal hydrolysis	Neyens and Baeyens (2003)
	• Slime-EPS content is greater than either loosely or tightly bound ECP for hydrolyzed digested sludge and increased significantly with reaction temperature between 80°C and 180°C for post digestion application of thermal hydrolysis.	Yang *et al.* (2019)
Foam and scum control	• A pilot-scale plant was successful in treating scum comprised of *Gordonia*-type organisms	Jolis and Marneri (2006)
	• Thermal hydrolysis at 170°C successfully used to abate foaming issues in downstream pilot-scale digesters. Significant reduction in abundance of filamentous organisms *M. parvicella* noted, although in contrast to other measurements which showed higher foam potential and stability with thermal hydrolysis	Alfaro *et al.* (2014)
	• Heat treatment prior to anaerobic digestion removed foaming by destroying hydrophobic materials	Barjenbruch and Kopplow (2003)
	• At lab-scale digesters fed thermally hydrolyzed sludge foamed more than those that were not	Oosterhuis *et al.* (2014)
Food and other co-substrates	• Biogas yield from thermally hydrolyzed food-waste lower (5–10%) than un-processed material at loading rates between 2 and 6 kg VS/m^3 d	Tampio *et al.* (2014)
	• Digestability of both kitchen waste and vegetable fruit residues (measured by BMP spikes) both reduced after thermal hydrolysis (175°C/60 min) by 14 and 8%, respectively	Liu *et al.* (2012)
	• Food waste co-digestion increases color, refractory COD, DON and UVA_{254} compared to municipal sludge	
Hydrolysis	• Rate increases with decreasing particle size	Aldin (2010)
	• Bacterial trophic groups specializing in hydrolysis largely absent on full-scale plants with thermal hydrolysis	Ennouri *et al.* 2016

Inhibition	• Inhibition in downstream digestion above 175°C	Stuckey and McCarty (1984) Haug *et al.* (1978)
	• No inhibition observed with digester feed of 10.4% DS	Oosterhuis *et al.* (2014)
	• No inhibition observed with digester feed up to 13% DS	Chauzy *et al.* (2008)
Loading rate for digestion	• Thermal hydrolysis increases loading rate by a factor of 2.3 compared to standard anaerobic digestion	Oosterhuis *et al.* (2014)
	• Typical design loading rate in the range of 5–6 kg VS/m^3 d	
	• However, loading rates up to 8 kg VS/m^3 d possible dependent on sludge type	
	• Loading rate of 11 kgCOD/m^3/d possible with granular expanded bed anaerobic digester	Qiao *et al.* (2013)
Odors and bacterial regrowth during dewatering	• Odorous compounds normally associated with heat treatment are significantly reduced during digestion of thermally pre-treated sludge.	Neyens and Baeyens (2003)
	• Significantly lowest odor potential during dewatering for both belt presses and high-speed centrifuges	Chen *et al.* (2011)
Particle size	• Increase in median size after thermal pre-treatment	Bougrier *et al.* (2006)
	• Decrease in average size from 70 to 35 µm	Barber (2010)
	• Decrease in average size from 107 to 66 µm	Neyens *et al.* (2004)
pH	• pH of hydrolyzed material decreases in a linear fashion between reaction temperatures of 100 and 220°C	Jeong *et al.* (2019)
	• pH drops from 7.5 to 6	Chen *et al.* (2018)
	• pH drops from 7.1 to 5.4	Han *et al.* (2017)
Phosphorous	• Little influence observed at full-scale	Tong *et al.* (2019)
	• Increase in dissolved organic phosphorous from 0.07 to 0.11 g/l after hydrolysis at 165°C for 50 min	Han *et al.* (2017)
	• Increase in phosphate from 0.05 to 0.7 g/l	Han *et al.* (2017)
	• Decrease in particulate phosphorous from 1.94 to 1.35 g/l	Han *et al.* (2017)
Polymer demand	• Increased by thermal hydrolysis at lab-scale from 10.5 to 18 kg active polymer/TDS	Oosterhuis *et al.* (2014)
	• However, no difference noted at full-scale, with dose of 8 kg active polymer/TDS for centrifuge dewatering.	Lancaster (2015)
	• Polymer demand for thickening $<$6 kg active polymer/TDS	
Pressure drop	• Release of both carbohydrates and proteins increases in an approximately linear fashion with increasing pressure drop between 3 and 6 bar. Release of lipids was uninfluenced by pressure drop.	Perrault *et al.* (2015)

(Continued)

Table 1.2 Influence of thermal hydrolysis (*Continued*).

Parameter	Observation	Reference
	• Biogas production rate is quicker with increasing pressure drop	Perrault *et al.* (2015)
	• Increasing flash pressure from 1.7 bar to 3 bar did not have a significant impact on the return liquor. However, increasing the flash pressure to 5 bar increased the DON and UV_{254} absorbing compounds.	Ahuja (2015)
Refractory material	• Solubility of carbohydrates reduces at higher temperatures hypothesized to be due to conversion to refractory material	Bougrier *et al.* (2008)
	• Equivalent to 11 kg COD/t DS dewatered sludge	Oosterhuis *et al.* (2014)
	• Thermal hydrolysis of both kitchen waste and vegetable fruit waste increases production of high molecular weight Maillard reaction products in response to increased solubility of both sugars and proteins	Liu *et al.* (2012)
Settleability	• Settleability (measured by SSVI) significantly improves with increasing temperature	Bougrier *et al.* (2008)
Solubility of carbohydrates	• Increase in solubility between 130°C and 210°C but little influence between 95°C and 130°C.	Bougrier *et al.* (2008)
	• Solubility of carbohydrates occurs below 150°C above which solubility of proteins becomes more evident.	Lu *et al.* (2014)
	• Increasing solubility with temperature between 130°C and 170°C. Strong linear increase between 10 and 30 min, after which further increases are negligible	Liu *et al.* (2012)
	• Increases with temperature between 125°C and 175°C	Noike *et al.* 1985)
	• Increases from approximately 500 mg/l to between 1,000 and 2,500 mg/l at two full-scale plants	Tong *et al.* (2019)
	• Increase from 0.5 to 3.97 g/l at 160°C for 30 min	Chen *et al.* (2018)
	• Minimal impact on solubility of carbohydrates	Han *et al.* (2017)
Solubility of COD	• Increases in a linear-type fashion with both temperatures (130 and 170°C) and reaction time (10–60 min)	Lu *et al.* (2014)
	• Solubility increases between 150°C and 180°C and reaction time (0–90 min). However, at 170°C and above no further solubilization between 60 and 90 min	Evert (1972)
	• For temperatures lower than 200°C, COD solubilization was found to increase linearly with treatment temperature for different sludge samples tested.	Bougrier and Carrere (2007)
	• Increases with temperature between 125°C and 175°C	Liu *et al.* (2012)
	• Increases from 12,000 and 5,000 mg/l to approximately 20,000 mg/l at two full-scale plants	Tong *et al.* (2019)
	• Vast majority of soluble COD is attributed to proteins which contribute around 55–60%, followed by lipids (15–25%) and polysaccharides (13–15%).	Donoso-Bravo *et al.* (2011)
		Donoso-Bravo *et al.* (2011)

	• The concentrations of soluble proteins and soluble polysaccharides are all strongly affected by reaction temperature at 150°C and above	Qiao *et al.* (2013)
	• Solubility increases with increasing reaction time from 30 to 60 to 90 min	Yang *et al.* (2019)
	• Soluble COD increased from approximately 3,000 to 30,000 mg/l with reaction temperature of 160°C for half hour retention time	Yang *et al.* (2019)
	• Soluble COD increased in a linear fashion from 5,000 and 20,000 between 80 and 180°C for previously digested sludge. Digested sludge control at mesophilic temperature had circa 1,500 mg/l	Jeong *et al.* (2019)
	• Increases in linear fashion between 100°C and 200°C reaction temperature. Further smaller increase at 220°C	Wang *et al.* (2009)
	• Large increase from 1.8 to 13.8 g/l at 170°C for 30 min	Chen *et al.* (2018)
	• Increase from approximately 7 to 47 g/l at 160°C for 30 min	Zhang *et al.* (2017)
	• Rheological measurements can be used as indicators to determine extent of solubilization with thermal hydrolysis	Hii *et al.* (2017)
	• Most solubility of COD within 20 minutes at temperatures between 100°C and 145°C	
Solubility of lipids	• Little influence from either temperature or reaction time.	Bougrier *et al.* (2008)
	• Negligible increases with temperature and reaction time.	Lu *et al.* (2014)
	• No influence	Li and Noike (1992), Wilson and Novak (2009)
Solubility of proteins	• Increases in a linear-type fashion with both temperature (130°C and 170°C) and reaction time (10–60 min)	Lu *et al.* (2014)
	• Increases with temperature between 125°C and 175°C	Liu *et al.* (2012)
	• Increases from approximately 1,000 to 7,500 mg/l at two full-scale plants	Tong *et al.* (2019)
	• Rate of protein solubilization far faster than polysaccharide solubilization. Both showed linear increases with temperatures between 60°C and 90°C and 120–180°C	Zhang *et al.* (2017)
	• Most solubilization of COD due to enhanced protein hydrolysis.	Han *et al.* (2017)
Suspended solids	• Reduction from 42 to 24 g/l at 170°C for 30 min	Wang *et al.* (2009)
UVA_{254}	• Increase in absorbance with increasing operating temperature. Non-linear which becomes far more pronounced between 150°C and 170°C	Ahuja (2015)
	• Increase in absorbance with increasing temperature. Effluent absorbance shows similar trend to that observed after thermal hydrolysis. Digestion removes absorbance	Higgins *et al.* (2015)
	• Increase in absorbance with increasing temperature	Dwyer *et al.* (2008)
	• Absorbance increases with increasing levels of activated sludge	

(Continued)

Table 1.2 Influence of thermal hydrolysis (*Continued*).

Parameter	Observation	Reference
VFAs	• Increase in levels of acetic and propionic acids with reaction time, little influence on C4 and above acids.	Lu *et al.* (2014)
	• VFA release from primary hydrolyzed sludge between 4 and 7 times higher than from activated sludge due to hydrolysis of unsaturated lipids	Wilson and Novak (2009)
	• Reduction in quantity of VFAs, similar composition	Tong *et al.* (2019)
	• Levels of VFAs increase exponentially with reactor temperature between 100°C and 220°C. Acetic acid contributes 75–80% of total VFAs	Jeong *et al.* (2019)
	• Investigating temperatures between 150°C and 240°C, highest VFA yield was at 190°C. Concentration of 5.38 g/l equivalent to 1,076.6 mg VFA/g VS_{added}. Most of VFAs were acetic acid	Iaqa Kakar *et al.* (2019)
	• Increase from 780 to 1,580 mg/l at 170°C for 30 min	Wang *et al.* (2009)
	• Increase from 1.17 to 2.81 g/l at 160°C for 30 min	Chen *et al.* (2018)
	• Increase from 0.25 to 4.21 g/l at 165°C for 50 min	Han *et al.* (2017)
Volatile sulphfur compounds in biogas	• Reduction of approximately 90% with thermal hydrolysis at laboratory scale	Wilson *et al.* (2011)
	• Dissolved sulfide increased from 1.9 to 5.5 g/l	Han *et al.* (2017)
Viscosity	• Falls significantly with increasing temperature between 130°C and 170°C	Higgins *et al.* (2015)
	• Significant reduction	Oosterhuis *et al.* (2014)
	• Following thermal hydrolysis (175°C for 60 min), viscosities of WAS, kitchen waste and vegetable and fruit residues reduced from 13,500; 36,000; and 6,250 to 1,625; 1,658 and 663 mPa s, respectively.	Liu *et al.* (2012)
	• Viscosity reduces further with thermal pre-treatment than it does with either ozonation or ultrasonic treatment.	Bougrier *et al.* (2006)
	• Thermal treatment of sludge (170°C and 190°C) changes rheological properties of WAS from non- to pseudo-Newtonian based on use of Ostwald de Waele relationship • Although thermal hydrolysis reduces apparent sludge viscosity by orders of magnitude, the viscosity of digested sludge is independent of the presence or absence of pre-treatment	Dawson and Ozgencil (2009)
	• Linear relationship between reduction in dimensionless yield stress, and also high shear viscosity (at 600 /s) with increasing solubilization of COD measured at temperatures between 100°C and 145°C for thermal hydrolysis reaction time of 60 min	Hii *et al.* (2017)
	• Most reduction in yield stress and apparent viscosity occurs within 10 min of heat treatment between 100°C and 145°C	Hii *et al.* (2017)
	• The viscosity and yield stress of thermally hydrolyzed sludge which has been cooled to ambient temperature compared to untreated sludge is dependent on thermal hydrolysis reaction temperature with reductions of approximately 28%, 38%, 53%, 70% and 89% corresponding to reaction temperatures of 80°C, 100°C, 120°C, 130°C and 145°C	Hii *et al.* (2017)

Figure 1.6 Thermal hydrolysis facility in Hias, Norway (Photo courtesy of Cambi).

1.1.5 Commercial growth of the technology

As previously mentioned, the first commercially operating plant was in Hias in Norway (Figure 1.6) in 1995, supplied by the market leader, Cambi based in Norway.

Figure 1.7 shows the uptake of commercial-scale facilities.

Between 1995 and 2000 most plants were in Scandinavia due to the proximity of the supplier Cambi. After the year 2000, some early adopters appeared in the UK, however, growth stagnated due to teething issues on some of the early facilities which deterred potential buyers. Initial issues were addressed in the mid-2000s, and combined with incentives for renewable energy generation, stimulated growth in the UK. At the time, UK Water Utilities, afraid of losing revenue potential from Renewable Obligation Certificates invested in technologies to enhance biogas production, and it was at this time that a second supplier, Veolia arrived with a technology called Thelys (subsequently Biothelys). The UK remains an important market for thermal hydrolysis and accounts for a third of global installed capacity.

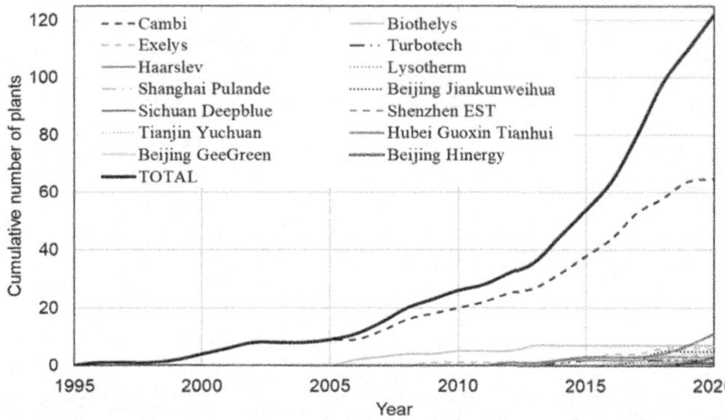

Figure 1.7 Commercial-scale growth of thermal hydrolysis in the municipal market (Data compiled from various sources. Data on China courtesy of Di Deng).

Table 1.3 Summary of global capacity for thermal hydrolysis in order of installed capacity*.

	Number	% of Total	Installed Capacity	% Installed	Largest	Smallest	Average	Median	Above 100	50–100	25–50	10–25	<10
Cambi	65	53	6947	77	450	4	107	77	23	21	11	9	1
Shanghai Pulande	6	5	432	5	200	20	72	41	2	1	3		
Sichuan Deepblue	11	9	324	4	60	10	29	24		2	2	7	
Tianjin Yuchuan	4	3	220	2	100	20	55	50		2	1	1	
Biothelys	7	6	206	2	90	3	19	5		2	1		4
Beijing Jiankunweihua	5	4	185	2	90	18	37	22		1	1	3	
Hubei Guoxin Tianhui	3	2	180	2	90	40	60	50		1	2		
Beijing GeeGreen	4	3	146	2	60	16	37	35		2		2	
Exelys	6	5	126	1	68	8	25	14		1	2	1	2
Shenzhen EST	3	2	70	1	40	10	23	20			1	2	
Turbotech	2	2	55	1	36	19	27	27			1	1	
Beijing Hinergy	1	1	40	0	40	40	40	40			1		
Lysotherm	2	2	36	0	33	4	18	18			1		1
Haarslev	3	2	32	0	17	5	11	10				1	2
Average					**98**	**15**							
Total	**122**		**8999**						**25**	**33**	**27**	**27**	**10**
% of Total									**20**	**27**	**22**	**22**	**8**

*Data on China, courtesy of Di Deng.

As UK Water Companies are privately owned, they have to accommodate the needs of their shareholders. In combination with financial regulations from the UK Water Regulator (OFWAT), this influences decision making to have a strong economic focus. Subsequently, most UK utilities are repeat purchasers of the technology to address strategic level biosolids management. Larger facilities, such as Davyhulme and later Blue Plains in the US (both described in Chapter 6) have stimulated growth for larger facilities across the globe resulting in five of the top 10 largest facilities in the world being installed in Beijing (also in Chapter 6). This success has spouted more suppliers recently as shown in Figure 1.7. There are further suppliers of the technology, however, these have no commercially available facilities to their name at present and are therefore not mentioned here. Table 1.3 summarizes known full-scale suppliers worldwide.

At the time of writing, there is growing interest in North America and the Asia-Pacific region. More recently, there has been a trend towards different configurations of thermal hydrolysis such as partial sludge treatment, thermal hydrolysis of digested sludge or downstream of digestion. These options are discussed in further detail in Chapter 2.

1.1.6 Summary

Thermal hydrolysis has evolved from a technology to enhance dewatering in the 1940s to one which has become a standardized unit operation within a sludge processing line with multiple, largely positive, influences on both sludge and wastewater operations (discussed further in Chapters 3 and 4). It has evolved, along with other "hydrolysis" technologies due to the invention of the activated sludge process, lack of innovation in digestion design and long life-time of anaerobic digestion assets. Early research has shown it to be most influential with the biodegradability and dewaterability of activated sludge rather than primary sludge. Materials with large quantities of protein and cellulose are very amenable to processing whereas those high in lipids and grease are scarcely impacted. Researchers have found that there is a temperature range where benefits are optimal. Lower temperatures result in less solubilization and release of biodegradable material, while higher temperatures cause the production of refractory materials which diminish biodegradability and can cause issues with wastewater effluent if not addressed. Traditionally, thermal hydrolysis has been incorporated prior to digestion on mixed sludge, however, newer configurations which process only a fraction of the sludge upstream, or even digested biosolids downstream have evolved. These configurations have addressed concerns with energy balance (discussed in Chapter 2) while making plants smaller. Chapter 8 discusses ideas on where the technology may lead to in the future.

REFERENCES

Ahuja N. (2015). Impact of Operating Conditions on Thermal Hydrolysis Pre-Treated Digestion Return Liquor. Thesis submitted to the faculty of the Virginia Polytechnic

Institute and State University in partial fulfillment of the requirements for the degree of Master of Science in Civil and Environmental Engineering.

Aldin S. (2010). The Effect of Particle Size on Hydrolysis and Modeling of Anaerobic Digestion. PhD thesis, Chemical and Biochemical Engineering Department, University of Western Ontario.

Alfaro N., Cano R. and Fdz-Polanco F. (2014). Effect of thermal hydrolysis and ultrasounds pretreatments on foaming in anaerobic digesters. *Bioresource Technology*, **170**, 477–482.

Ardern E. and Lockett W. T. (1914). Experiments on the oxidation of sewage without the aid of filters. *Journal of the Society of Chemical Industry*, **33**(10), 523–539.

Asian Development Bank. (2012). Promoting Beneficial Sewage Sludge Utilization in the People's Republic of China.

Auty D. and Blake J. (2009). Strategic Carbon Technology – How To Use Sludge to Your Advantage. *13th European Biosolids & Organic Resources Conference & Workshop.*

Barber W. P. F. (2008). Modeling the carbon footprint of various biosolids treatment options. *Proceedings of the Water Environment Federation*, **2008**(7), 7787–7807.

Barber W. P. F. (2010). The influence on digestion and advanced digestion on the environmental impacts of incinerating sewage sludge – a case study from the UK. *Proceedings of the Water Environment Federation*, **2010**(4), 865–881.

Barber W. P. F. (2016). Thermal hydrolysis for sewage treatment: a critical review. *Water Research*, **104**, 53–71.

Barber W. P. F. (2017). Thermal hydrolysis reduces carbon footprint of biosolids treatment. *Proceedings of the Water Environment Federation*, **2017**(1), 85–105.

Barjenbruch M. and Kopplow O. (2003). Enzymatic, mechanical and thermal pre-treatment of surplus sludge. *Advances in Environmental Research*, **7**(3), 715–720.

Barlindhaug J. and Ødegaard H. (1996). Thermal hydrolysis for the production of carbon source for denitrification. *Water Science and Technology*, **34**(1), 371–378.

Batstone D. J., Jensen P. D. and Ge H. (2011). Biochemical treatment of biosolids-emerging technologies. *Water*, **5**, 90–93.

Bi X., Liu C., Ran X., Lv W., Wang J. and Liao Z. (2013). Experimental Studies on Pre-Treatment of Sludge by Combination of Thermal Hydrolysis and Ultrasound and Sand Removal by Hydrocyclone to Optimize Wastewater and Sludge Treatment. *Proceedings of Aquaenviro's 18th European Biosolids and Organic Resources Conference and Exhibition*, Manchester, UK.

Bougrier C., Albasi C., Delgenès J. P. and Carrère H. (2006). Effect of ultrasonic, thermal and ozone pre-treatments on waste activated sludge solubilisation and anaerobic biodegradability. *Chemical Engineering and Processing: Process Intensification*, **45**(8), 711–718.

Bougrier C., Delgenès J. P. and Carrère H. (2008). Effects of thermal treatments on five different waste activated sludge samples solubilisation, physical properties and anaerobic digestion. *Chemical Engineering Journal*, **139**(2), 236–244.

Bougrier D. J. and Carrere H. (2007). Impacts of thermal pre-treatments on the semi-continuous anaerobic digestion of waste activated sludge. *Biochemical Engineering Journal* **34**(1), 20–27.

Brooks R. B. (1968). Heat treatment of activated sludge. *Water Pollution Control*, **67**, 592–601.

Brooks R. B. (1970). Heat treatment of sewage sludge. *Water Pollution Control*, **69**, 92–99.

Buswell A. M. and Neave S. L. (1930). Laboratory studies of sludge digestion. Bulletin (Illinois State Water Survey) no. 30.

Chauzy J., Cretenot D., Bausseron A. and Deleris S. (2008). Anaerobic digestion enhanced by thermal hydrolysis: first reference BIOTHELYS® at Saumur, France. *Water Practice and Technology*, **3**(1), wpt2008004.

Chen S., Ning L., Dong B., Zhao W., Dai L. and Dai X. (2018). New insights into the enhanced performance of high solid anaerobic digestion with dewatered sludge by thermal hydrolysis: organic matter degradation and methanogenic pathways. *Journal of Hazardous Materials*, **342**, 1–9.

Chen Y. C., Higgins M. J., Beightol S. M., Murthy S. N. and Toffey W. E. (2011). Anaerobically digested biosolids odor generation and pathogen indicator regrowth after dewatering. *Water Research*, **45**(8), 2616–2626.

Chu C. P., Lin W. W., Lee D. J., Chang B. V. and Peng X. F. (2002). Thermal treatment of waste activated sludge using liquid boiling. *Journal of environmental engineering*, **128**(11), 1100–1103.

Dawson M. K. and Ozgencil H. (2009). Do Hydrolysis Processes Effect Digested Sludge Rheology? *Proceedings of Aquaenvio 14th European Biosolids and Organic Resources Conference and Exhibition*, Leeds, UK.

Donoso-Bravo A., Pérez-Elvira S., Aymerich E. and Fdz-Polanco F. (2011). Assessment of the influence of thermal pre-treatment time on the macromolecular composition and anaerobic biodegradability of sewage sludge. *Bioresource technology*, **102**(2), 660–666.

Dwyer J., Starrenburg D., Tait S., Barr K., Batstone D. J. and Lant P. (2008). Decreasing activated sludge thermal hydrolysis temperature reduces product colour, without decreasing degradability. *Water Research*, **42**(18), 4699–4709.

Ennouri H., Miladi B., Diaz S. Z., Güelfo L. A. F., Solera R., Hamdi M. and Bouallagui H. (2016). Effect of thermal pretreatment on the biogas production and microbial communities balance during anaerobic digestion of urban and industrial waste activated sludge. *Bioresource Technology*, **214**, 184–191.

Evert J. G. (1972). Dewatering of wastewater sludge by heat treatment. *Journal Water Pollution Control Federation*, **44**(1), 92–100.

Goodwin J. A. S. and Forster C. F. (1985). A further examination into the composition of activated sludge surfaces in relation to their settlement characteristics. *Water Research*, **19**(4), 527–533.

Gossett J. M. and Belser R. L. (1982). Anaerobic digestion of waste activated sludge. *Journal of the Environmental Engineering Division*, **108**(6), 1101–1120.

Han Y., Zhuo Y., Peng D., Yao Q., Li H. and Qu Q. (2017). Influence of thermal hydrolysis pretreatment on organic transformation characteristics of high solid anaerobic digestion. *Bioresource Technology*, **244**, 836–843.

Haug R. T. (1977). Sludge processing to optimize digestibility and energy production. *Journal (Water Pollution Control Federation)*, 1713–1721.

Haug R. T., Stuckey D. C., Gossett J. M. and McCarty P. L. (1978). Effect of thermal pretreatment on digestibility and dewaterability of organic sludges. *Journal Water Pollution Control Federation*, **50**(1), 73–85.

Higgins M. J., Beightol S., Mandahar U., Suzuki R., Xiao S., Lu H. W., ... and Novak J. T. (2017). Pretreatment of a primary and secondary sludge blend at different thermal hydrolysis temperatures: impacts on anaerobic digestion, dewatering and filtrate characteristics. *Water Research*, **122**, 557–569.

Hii K., Parthasarathy R., Baroutian S., Gapes D. J. and Eshtiaghi N. (2017). Rheological measurements as a tool for monitoring the performance of high pressure and high temperature treatment of sewage sludge. *Water Research*, **114**, 254–263.

Hiraoka M., Takeda N., Sakai S. L. and Yasuda A. (1985). Highly efficient anaerobic digestion with thermal pre-treatment. *Water Science & Technology*, **17**(4), 529–539.

Jeong S. Y., Chang S. W., Ngo H. H., Guo W., Nghiem L. D., Banu J. R., ... and Nguyen D. D. (2019). Influence of thermal hydrolysis pretreatment on physicochemical properties and anaerobic biodegradability of waste activated sludge with different solids content. *Waste Management*, **85**, 214–221.

Jolis D. and Marneri M. (2006). Thermal hydrolysis of secondary scum for control of biological foam. *Water Environment Research*, **78**(8), 835–841.

Lancaster R. (2015). Thermal Hydrolysis at Davyhulme WWtW – One Year. *Proceedings of WEF Residuals and Biosolids 2015, Water and Environment Federation*, Washington DC.

laqa Kakar F., Koupaie E. H., Razavi A. S., Hafez H. and Elbeshbishy E. (2019). Effect of hydrothermal pretreatment on volatile fatty acids production from thickened waste activated sludge. *BioEnergy Research*, 1–14.

Li Y. Y. and Noike T. (1992). Upgrading of anaerobic digestion of waste activated sludge by thermal pretreatment. *Water Science & Technology*, **26**(3–4), 857–866.

Li X. Y. and Yang S. F. (2007). Influence of loosely bound extracellular polymeric substances (EPS) on the flocculation, sedimentation and dewaterability of activated sludge. *Water Research*, **41**, 1022–1030.

Liu X., Wang W., Ga X., Zhou Y. and Shen R. (2012). Effect of thermal pretreatment on the physical and chemical properties of municipal biomass waste. *Waste Management*, **32**(2), 249–255.

Lu H. W., Xiao S., Le T., Al-Omari A., Higgins M., Boardman G., Novak J. and Murthy S. (2014). Evaluation of Solubilization Characteristics of Thermal Hydrolysis Process. *Proceedings of the Water Environment Federation*, **2014**(15), 6312–6336.

Lumb C. (1940). Heat treatment as an aid to sludge dewatering-ten months' full-scale operation. *Surveyor*, **98**(2550), 287–291.

Lumb C. (1951). Heat treatment as an aid to sludge dewatering: ten years' full-scale operation. *J. Proc. Inst. Sew. Purif*, **198**(2550), 287–291.

Mohlman F. W. (1917). The activated-sludge method of sewage treatment. *University of Illinois Bulletin*, **XV**(11), 75–114.

Netteland T., Sørensen G. and Solheim O. E. (2015). 20 Years of Continuous Operating of the World's First Thermal Hydrolysis Plant at Hias IKS. *Lessons Learned Proceedings of Aquaenviro 20th European Biosolids Conference*, Leeds, UK.

Neyens E. and Baeyens J. (2003). A review of thermal sludge pre-treatment processes to improve dewaterability. *Journal of Hazardous Materials*, **98**(1), 51–67.

Neyens E., Baeyens J. and Dewil R. (2004). Advanced sludge treatment affects extracellular polymeric substances to improve activated sludge dewatering. *Journal of Hazardous Materials*, **106**(2–3), 83–92.

Noike T., Endo G., Chang J. E., Yaguchi J. I. and Matsumoto J. (1985). Characteristics of carbohydrate degradation and the rate-limiting step in anaerobic digestion. *Biotechnology and Bioengineering*, **27**(10), 1482–1489.

Ødeby T., Netteland T. and Solheim O. E. (1996). Thermal hydrolysis as a profitable way of handling sludge. In *Chemical Water and Wastewater Treatment IV*, Springer, Berlin, Heidelberg, 401–409.

Oosterhuis M., Ringoot D., Hendriks A. and Roeleveld P. (2014). Thermal hydrolysis of waste activated sludge at Hengelo wastewater treatment plant. The Netherlands. *Water Science and Technology*, **70**(1), 1–7.

Ostapczuk R. E., Bassette P. C., Dassanayake C. and Bevington G. (2011). Recuperative thickening: decoupling the SRT from the HRT reduces capital expenditures and increases biogas production for CHP utilization. *Proceedings of the Water Environment Federation*, **2011**(15), 2348–2355.

Perrault A., Thorpe R. B., Cataldi A., Mills N. and Barua R. (2015). Steam Explosion of Sludge in the THP Process. *Proceedings of Sludgetech Conference*, Guilford, Surry, UK.

Phothilangka P., Schoen M. A. and Wett B. (2008). Benefits and drawbacks of thermal pre-hydrolysis for operational performance of wastewater treatment plants. *Water Science and Technology*, **58**(8), 1547.

Pinnekamp J. (1989). Effects of thermal pretreatment of sewage sludge on anaerobic digestion. *Water Science and Technology*, **21**(4–5), 97–108.

Qiao W., Yin Z., Wang W., Wang J. and Zhang Z. (2013). Pilot-scale experiment on thermally hydrolyzed sludge liquor anaerobic digestion using a mesophilic expanded granular sludge bed reactor. *Water Science and Technology*, **68**(4), 948–955.

Scott-Moncrieff W. D. (1898). The anaërobic bacterial treatment of sewage. *British Medical Journal*, **2**(1964), 514.

Sherwood R. and Phillips J. (1970). Heat treatment process improves economics of sludge handling and disposal. *Water and Wastes Engineering*, **7**(11), 42–44.

Stuckey D. C. and McCarty P. L. (1984). The effect of thermal pretreatment on the anaerobic biodegradability and toxicity of waste activated sludge. *Water Research*, **18**(11), 1343–1353.

Svennevik O. K., Solheim O. E., Beck G., Sørland G. H., Jonassen K. R., Rus E., Westereng B., Horn S. J., Higgins M. J. and Nilsen P. J. (2019). Effects of post anaerobic digestion thermal hydrolysis on dewaterability and moisture distribution in digestates. *Water Science and Technology*, **80**(7), 1338–1346.

Tampio E., Ervasti S., Paavola T., Heaven S., Banks C. and Rintala J. (2014). Anaerobic digestion of autoclaved and untreated food waste. *Waste Management*, **34**(2), 370–377.

Tong J., Fang P., Zhang J., Wei Y., Su Y. and Zhang Y. (2019). Microbial community evolution and fate of antibiotic resistance genes during sludge treatment in two full-scale anaerobic digestion plants with thermal hydrolysis pretreatment. *Bioresource Echnology*, **288**, 121575.

Torpey W. N. and Melbinger N. R. (1967). Reduction of digested sludge volume by controlled recirculation. *Journal (Water Pollution Control Federation)*, **39**(9), 1464–1474.

Urbain V., Block J. C. and Manem J. (1993). Bioflocculation in activated sludge: an analytic approach. *Water Research*, **27**(5), 829–838.

van Dijk L. and de Man A. (2010). TurboTec®: continue thermische slib hydrolyse voor lagere zuiveringskosten en meer energie op de rwzi. *Neerslag* **2010**, 45, 25–33.

Wang L., Chang Y. and Li A. (2019). Hydrothermal carbonization for energy-efficient processing of sewage sludge: a review. *Renewable and Sustainable Energy Reviews*, **108**, 423–440.

Wang Z., Wang W., Zhang X. and Zhang G. (2009). Digestion of thermally hydrolyzed sewage sludge by anaerobic sequencing batch reactor. *Journal of Hazardous Materials*, **162**(2–3), 799–803.

Willis J., Bailey W., Jolly E., Murthy S., Peot C. and Trueblood D. (2010). DCWASA'S certifiable GHG inventory and projected GHG reductions from Cambi/anaerobic digestion upgrades. *Proceedings of the Water Environment Federation*, **2010**(2), 141–147.

Wilson C. A. and Novak J. T. (2009). Hydrolysis of macromolecular components of primary and secondary wastewater sludge by thermal hydrolytic pretreatment. *Water Research*, **43**(18), 4489–4498.

Wilson C. A., Tanneru C. T., Banjade S., Murthy S. N. and Novak J. T. (2011). Anaerobic digestion of raw and thermally hydrolyzed wastewater solids under various operational conditions. *Water Environment Research*, **83**(9), 815–825.

Yang D., Hu C., Dai L., Liu Z., Dong B. and Dai X. (2019). Post-thermal hydrolysis and centrate recirculation for enhancing anaerobic digestion of sewage sludge. *Waste Management*, **92**, 39–48.

Yang S., Nghiem L. D., Bustamante H., van Rys D. and Murthy S. N. (2015). Recuperative thickening: a possible tool to improve anaerobic digestion of wastewater sludge. *OZWater*.

Yang S., Phan H. V., Bustamante H., Guo W., Ngo H. H. and Nghiem L. D. (2017). Effects of shearing on biogas production and microbial community structure during anaerobic digestion with recuperative thickening. *Bioresource Technology*, **234**, 439–447.

Zack S. I. and Edwards G. P. (1929). Gas production from sewage sludge. *Sewage Works J.*, **1**(2), 160–186.

Zhang J., Xue Y., Eshtiaghi N., Dai X., Tao W. and Li Z. (2017). Evaluation of thermal hydrolysis efficiency of mechanically dewatered sewage sludge via rheological measurement. *Water Research*, **116**, 34–43.

Zhen G., Lu X., Kato H., Zhao Y. and Li Y. Y. (2017). Overview of pretreatment strategies for enhancing sewage sludge disintegration and subsequent anaerobic digestion: current advances, full-scale application and future perspectives. *Renewable and Sustainable Energy Reviews*, **69**, 559–577.

Chapter 2

Design – Mass and energy balance

"There are three important factors to be considered....as follows: 1. The specific heat of the sludge, 2. The heat conductivity of the sludge, and 3. The generation of heat by biological processes."

F. Sierp (translated by F. Mohlman) 1930

2.1 OVERVIEW

While thermal hydrolysis improves the biodegradability of sludge, and hence increases biogas production during anaerobic digestion, energy is needed to meet the reaction temperatures of the process. Although most of the energy required can be recovered by exploiting high-grade heat from co-generation, a small quantity of parasitic demand remains. The additional energy production from the extra biogas generated by employing the process is not generally a sufficient driver for installation of thermal hydrolysis. However, the energy requirement is fundamentally influenced by the dry solids (DS) content of the sludge being hydrolyzed decreasing with increasing dryness, and by the temperature differential needed. As will be shown, the sludge composition is also fundamentally important, and energy balance deteriorates with increasing levels of biological sludge. Optimizing these parameters can result in a system approaching one which is independent of external energy requirements. Another way to reduce or eliminate the energy demands of thermal hydrolysis is to treat less sludge. If there is no driver for pathogen-free digested biosolids, it is feasible to treat only a fraction of the sludge. Preferably, if primary and biologically

Author: William Barber
doi: 10.2166/9781789060287_0027

produced sludge are thickened separately, it is possible to focus on treatment of the biological sludge fraction as it is on those types of sludges where thermal hydrolysis has the greatest impact. Alternatively, if there is sufficient anaerobic digestion capacity, treatment of digested sludge prior to re-digestion remains an option.

2.1.1 Energy demand of thermal hydrolysis

A key factor regarding thermal hydrolysis is minimization of the energy requirement needed to reach reaction temperature. Subsequently, it is important to optimize the quantity and temperature of the sludge being processed. As the sludge moiety has a lower specific heat capacity than water (Xu and Lancaster, 2009), increasing the DS of the sludge will intrinsically reduce energy requirements. Typically, the sludge DS are thickened to between 15 and 30% DS depending on equipment supplier and configuration. While thickening sludge reduces energy demand, heat-transfer limitations, potential for short-circuiting and practical processing concerns such as whether it is even possible to thicken waste-activated sludge to high DS need to be considered. From a financial perspective, the cost of additional polymer required to achieve the higher DS is typically higher than the cost savings due to reduced energy (and therefore natural gas) requirements.

If one considers the thermal hydrolysis plant as a black box as shown in Figure 2.1, then the energy required for thermal hydrolysis may be calculated from elementary heat-transfer theory as follows:

$$Q = (m_s C_{ps} \Delta T_s + m_w C_{pw} \Delta T_w)/\eta \tag{2.1}$$

where

Q = energy required (kJ)
m = mass (kg)
C_p = specific heat capacity (kJ/kg/°C) = 1.5 kJ/kg/°C for sludge and 4.2 kJ/kg/°C
ΔT = temperature difference required (°C) = ($T_{out} - T_{in}$)
η = efficiency to account for heat losses by convection and radiation
Subscripts s and w refer to sludge and water

2.1.1.1 *Influence of thickening*

As mentioned, thickening of the sludge is key in reducing the overall energy demand of the system. The influence of DS content on the energy demand of thermal hydrolysis can be demonstrated by way of an example as follows:

Question. *Work out how much less energy is required to heat 1 t DS of sludge at 20% DS compared to 1 t DS of sludge at 16.5% DS? Assume specific heat capacities for sludge and wastewater are 1.5 and 4.2 kJ/kg/°C, respectively,*

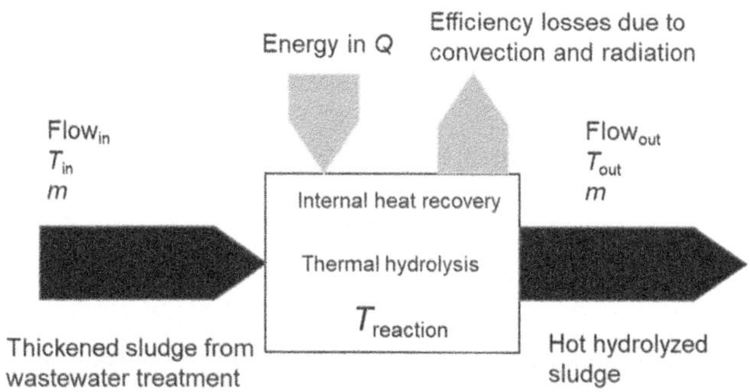

Figure 2.1 Thermal hydrolysis black box showing energy requirements.

temperature difference required is 75°C and system efficiency is 65%. Density of wastewater assumed as 1,000 kg/m³.

Answer.

1 t DS contains 1 × 1,000 kg = 1,000 kg dry material

At 20% DS, the wet weight of sludge is 1,000/20% = 5,000 kg of which 1,000 kg are dry and (5,000 – 1,000=) 4,000 kg are wet.

Energy to heat sludge

= 1,000 kg × 1.5 kJ/kg/°C × 75°C temperature increase = 112,500 kJ

Energy to heat water

= 4,000 kg × 4.2 kJ/kg/°C × 75°C temperature increase = 1,260,000 kJ

Therefore, energy to heat sludge at 20% DS before efficiency losses

= 112,500 + 1,260,000 = 1,372,500 kJ

Which after losses becomes 1,372,500/65% = 2,111,538 kJ, or 587 kWhr/t DS hydrolyzed.

Over 90% of the total energy required is due to the heating of water within the sludge and not the thickened solids, highlighting the importance of thickening.

Recalculating the above at 16.5% DS

1 t DS contains 1 × 1,000 kg = 1,000 kg dry material

At 16.5% DS, the wet weight of sludge is 1,000/16.5% = 6,061 kg of which 1,000 kg are dry and (6,061 – 1,000=) 5,061 are wet.

Energy to heat sludge = 112,500 kJ as before

> *Over 90% of the energy required for thermal hydrolysis is to heat water even if the sludge is thickened to 25% dry solids*

Energy to heat water

$$= 5{,}061 \text{ kg} \times 4.2 \text{ kJ/kg/°C} \times 75\text{°C temperature difference} = 1{,}594{,}091 \text{ kJ}$$

Therefore, energy to heat sludge at 16.5% DS before efficiency losses

$$= 112{,}500 + 1{,}594{,}091 = 1{,}706{,}591 \text{ kJ}$$

Which after losses becomes 1,706,591/65% = 2,625,524 kJ = 729 kWhr/t DS processed.

This is an increase in energy of 24% compared to heating the sludge to 20%.

2.1.1.2 *Influence of temperature difference*

However, the amount of heat recovered by the thermal hydrolysis plant is also a very significant factor. In the above example it is assumed that a temperature difference of 75°C is required. Typically, systems which process higher DS have no or negligible heat recovery resulting in a larger temperature differential. Assume the example with sludge thickened to 20% has no recovery of heat and sludge temperature entering thermal hydrolysis is 30°C and it exits at 165°C. How much energy is now required?

At 20% DS, the wet weight of sludge is 1,000/20% = 5,000 kg of which 1,000 kg are dry and (5,000 – 1,000=) 4,000 kg are wet.

Energy to heat sludge

$$= 1{,}000 \text{ kg} \times 1.5 \text{ kJ/kg/°C} \times (165\text{°C} - 30\text{°C}) \text{ temperature increase}$$

$$= 202{,}500 \text{ kJ}$$

Energy to heat water

$$= 4{,}000 \text{ kg} \times 4.2 \text{ kJ/kg/°C} \times (165\text{°C} - 30\text{°C}) \text{ temperature increase}$$

$$= 2{,}268{,}000 \text{ kJ}$$

Therefore, energy to heat sludge at 20% DS before efficiency losses

$$= 202{,}500 + 2{,}268{,}000 = 2{,}470{,}500 \text{ kJ}$$

After which losses become 2,470,500/65% = 3,800,769 kJ or 1,056 kWhr/t DS processed.

This system now requires 3,800,769 – 2,625,524 = 1,175,245 kJ more energy than the system running at 16.5% DS, even though the sludge is thicker. This is an increase of over 30% compared to the scenario at lower DS.

For this system to have the same energy requirement as the one processing 16.5% DS with the lower temperature difference, the sludge would have to be thickened to a minimum of 28% DS.

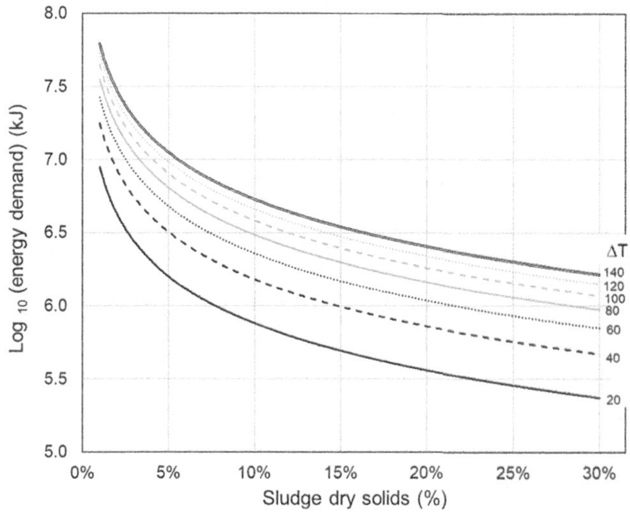

Figure 2.2 Influence on temperature difference (ΔT) in °C and sludge DS (%) on energy demand of thermal hydrolysis before losses.

Figure 2.2 shows the influence of thickening sludge and the temperature difference required on the energy demand of thermal hydrolysis.

From the graph the following equation can be extracted to approximate the thermal energy demand of thermal hydrolysis:

$$Q = (10^{(-0.464\mathrm{Ln}(D)+c)})/\eta \tag{2.2}$$

where

Q = energy required (kJ)
D = DS of sludge expressed as a percentage, and
c is a constant which is a function of temperature difference ΔT and given by

$$c = 3.8233(\Delta T)^{0.0786} \tag{2.3}$$

and η = efficiency as before.

Consider the above example where 1 t of dry sludge is thickened to 16.5% DS with a temperature difference of 75°C, calculate the energy demand for thermal hydrolysis, assuming efficiency is 65%.

Inserting into equation 2.2:

$$Q = (10^{(-0.464\mathrm{Ln}(0.15)+c)})/0.65$$
$$c = 3.8233(\Delta T)^{0.0786} = 3.8233(75°\mathrm{C})^{0.0786} = 5.368$$

substituting into the above:

$$Q = (10^{(-0.464\mathrm{Ln}(0.15)+5.368)})/0.65$$

$$Q = (10^{6.248})/0.65$$
$$Q = 2{,}723{,}244 \text{ kJ}$$

This figure is within 4% of that calculated earlier.

2.1.1.3 Steam requirement

The energy required to reach thermal hydrolysis reaction temperatures is typically provided by steam. The steam requirement is influenced by temperature difference and sludge thickness as shown above. The demand is simply the energy required divided by the specific enthalpy of saturated steam at the appropriate temperature and subsequent pressure. In this instance, at an operating temperature of 165°C, specific enthalpy is 2,763 kJ/kg from steam tables. Therefore, for this example based on heating sludge with a thickness of 16.5% DS the steam requirement is

$$2{,}625{,}524 \text{ (kJ)}/2{,}763 \text{ (kJ/kg)} = 950 \text{ kg steam/t DS}$$

The energy required to provide steam can be provided in numerous ways including: direct use of boilers running on either bio- or natural gas; co-generation using reciprocating internal combustion engine with auxiliary boiler running on bio- or natural gas; or use of gas turbine on larger facilities, although the latter comes with a loss of power generation unless operated in combined cycle (Fernández-Polanco and Tatsumi, 2016). The quantity of energy required is typically described as a fraction of the biogas generated, however, comparing literature data reveals that the energy required does not follow a predictable pattern (Lancaster, 2015; Merry and Oliver, 2015; Pook *et al.*, 2013). This is due to several parameters which include: type, efficiency, configuration and availability of co-generation plant; presence and quantity of gas storage; gas production profiles; configuration, operating temperature and retention time of anaerobic digestion plant, and by far the most important parameter – albeit habitually overlooked, the sludge composition itself. In order to elucidate the wide-ranging energy demands observed in the literature, it is necessary to look closer at the energy balance for a theoretical situation which is independent of the variability of full-scale installations.

2.1.2 Mass and energy balance

As mentioned in Chapter 1, thermal hydrolysis influences a number of parameters involving sludge treatment. It has to be thickened first to reduce the energy demand, it allows digesters to be fed at higher DS, it improves the degradation of sludge which increases biogas production, it improves dewaterability and can change polymer dose and finally it changes the composition of the liquid sludge exiting digestion. It is not possible to extract data from full-scale mass balances to determine the influence of thermal hydrolysis on these parameters due to numerous site-specific conditions such as: digester type; type of mixing; retention time; digester configuration; digester temperature; feeding regime; presence or absence of co-generation; availability of co-generation etc. Therefore, to demonstrate a typical mass balance, a calculation was performed based on a plant processing all of a 60:40 blend of primary and activated sludge with an annual quantity of 10,000 t, based on theoretical considerations and design criteria, and demonstrated in Figure 2.3.

Question. *For the above example how much of the energy required for thermal hydrolysis – based on a temperature difference of 75°C and operating with 65% efficiency – can be provided by high-grade heat recovered from a co-generation plant based on an engine which has 92% availability and converts the biogas energy into 38% electricity, 25% high-grade heat, 17% low-grade heat and the rest is lost? Assume: 10,000 tonnes dry solids per year processed, 75% volatile content; volatile solid destruction is 62 and 50% for primary and waste-activated sludge fractions, respectively; biogas yields are 1.1 and 0.8 Nm^3/kg VS destroyed for primary and waste-activated sludge, respectively; water at 25°C is added to dilute hydrolyzed sludge such that the digester is fed at 10% DS. Determine how much parasitic load is required after high-grade heat has been accounted for?*

Answer.
This can be broken down into three parts, firstly determination of the energy required for thermal hydrolysis, followed by calculation of biogas energy, and finally how much of that energy is available.

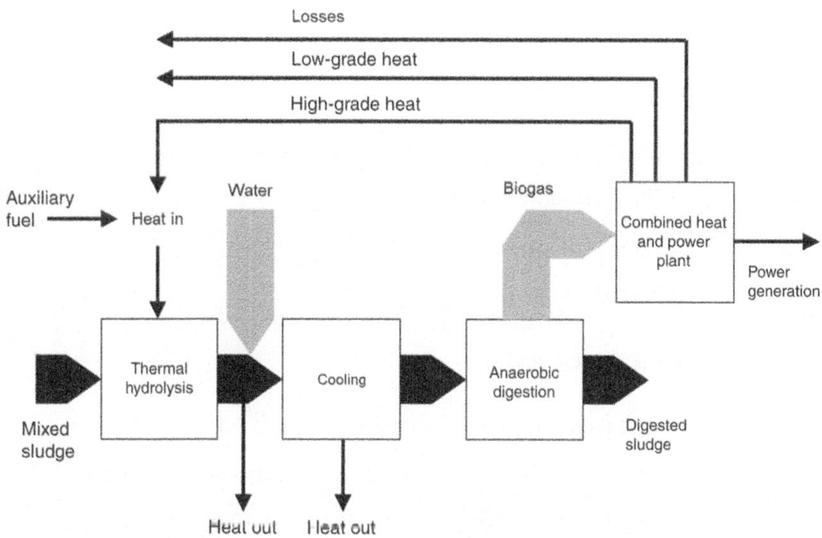

Figure 2.3 Basic energy demands of thermal hydrolysis.

(1) Energy for thermal hydrolysis

This is calculated as described before using equation 2.1.

At 16.5% DS, the wet weight of sludge is 1,000 (convert kg to tonne) × 10,000/16.5% = 60,606,060 kg of which 10,000,000 kg are dry and the remaining 50,606,060 are wet.

Energy to heat sludge

$$= 10{,}000{,}000 \text{ kg} \times 1.5 \text{ kJ/kg/°C} \times 75\text{°C temperature increase}$$
$$= 1{,}125{,}000{,}000 \text{ kJ} = 312{,}500 \text{ kWhr.}$$

From now energy figures will be converted to kWhr (1 kWhr = 1/3,600 kJ) for convenience

Energy to heat water

$$= 50{,}606{,}060 \text{ kg} \times 4.2 \text{ kJ/kg/°C} \times 75\text{°C temperature difference}$$
$$\times 1/3{,}600 \text{ (kJ to kWhr)}$$
$$= 4{,}428{,}030 \text{ kWhr.}$$

Therefore, energy to heat sludge at 16.5% after losses = 7,293,123 kWhr.

(2) Biogas energy

This is simply the product of the quantity of volatile material destroyed and the biogas yield.

For primary sludge

$$= 10{,}000 \text{ (toones/yr)} \times 1{,}000 \text{ (kg/tonne)} \times 60\% \text{ (primary sludge)}$$
$$\times 75\% \text{ (volatile fraction)}$$
$$\times 62\% \text{ (volatile solids destruction) } 1.1 \text{ (Nm}^3\text{/kgVS}_{\text{destroyed}})$$
$$= 3{,}069{,}000 \text{ Nm}^3 \text{ biogas/year}$$

For waste-activated sludge

$$= 10{,}000 \text{ (tonnes/yr)} \times 1{,}000 \text{ (kg/tonnes)}$$
$$\times 40\% \text{ (waste-activated sludge)} \times 75\% \text{ (volatile fraction)}$$
$$\times 50\% \text{ (volatile solids destruction)} \times 0.8 \text{ (Nm}^3\text{/kg VS}_{\text{destroyed}})$$
$$= 1{,}200{,}000 \text{ Nm}^3 \text{ biogas/year}$$

Therefore, total gas production $= 3{,}069{,}000 + 1{,}200{,}000 = 4{,}269{,}000\ \text{Nm}^3$

$$\begin{aligned}\text{Energy within gas} &= 4{,}269{,}000 \times 65\%\ (\text{methane by volume}) \\ &\quad \times 10.6\ (\text{kWhr/m}^3\ \text{methane}) \\ &= 29{,}413{,}410\ \text{kWhr.}\end{aligned}$$

Therefore, prior to energy recovery the thermal hydrolysis plant needs $7{,}293{,}123/29{,}413{,}410 = 24.8\%$ equivalence of the biogas energy.

(3) How much of the biogas energy is available?

The biogas energy goes into a co-generation unit where 25% is converted into high-grade heat and the engine is 92% available.

So, the high-grade heat available

$$\begin{aligned}&= 29{,}413{,}410\ \text{kWhr} \times 25\%\ (\text{conversion to high-grade heat}) \\ &\quad \times 92\%\ (\text{engine efficiency}) \\ &= 6{,}765{,}084\ \text{kWhr.}\end{aligned}$$

Therefore, parasitic load still required $= 7{,}293{,}123\ \text{kWhr} - 6{,}765{,}084\ \text{kWhr} = 528{,}039\ \text{kWhr}$ which is now equivalent to 1.8% biogas equivalence, or $528{,}039\ \text{kWhr}/10.6\ \text{kWhr/Nm}^3$ methane $= 49{,}815\ \text{Nm}^3$ methane annually before combustion losses.

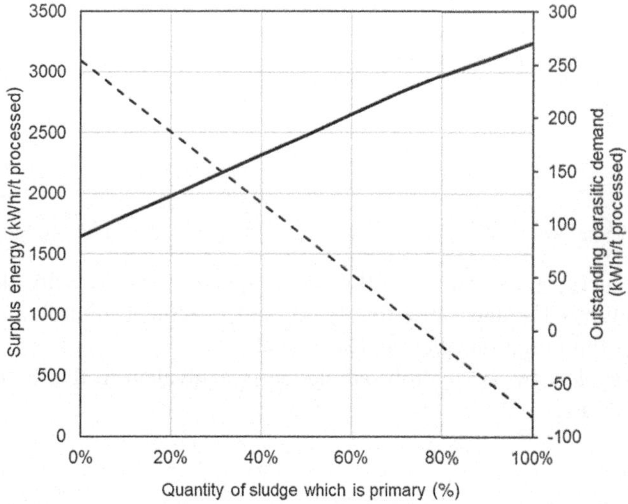

Figure 2.4 Influence of sludge composition on surplus energy produced (primary axis full line) and parasitic energy demand (secondary axis dashed line) after accounting for high-grade heat recovery from co-generation.

As well as DS and temperature difference across thermal hydrolysis as previously mentioned, the above calculations are critically dependent on sludge composition, as this influences the quantity of biogas (and hence amount of high-grade heat), and type and age of co-generation plant. Figure 2.4 is a sensitivity analysis on the above example where only the composition of sludge has been altered.

Systems with over 75% primary sludge are highly unlikely to have a parasitic energy demand even if all sludge is thermally hydrolyzed

As one would expect, as the quantity of primary sludge goes up, the surplus energy produced increases which reduces parasitic demand. This is ironic since thermal hydrolysis has greater influence on waste-activated sludge. The analysis shows no parasitic energy demand if primary sludge comprises over approximately 75% of the sludge being fed to the thermal hydrolysis unit. This composition is dependent on the assumptions made in the calculation with respect to engine efficiency and availability.

2.1.3 Cooling requirements

The hydrolyzed sludge has to be cooled to temperatures conducive with anaerobic digestion prior to being fed to the digestion plant. Fortunately, the sludge also requires dilution using pathogen-free water, final effluent or similar, in order to control the loading rate into the digester. This water is added prior to the cooling stage in order to reduce the size of the cooling unit required. To work out the cooler size, it is necessary to determine the temperature of the mixture of hydrolyzed sludge and dilution water. As before, this is determined by straightforward heat transfer where:

$$Q_{\mathrm{H}} = m_{\mathrm{H}} C_{p\mathrm{H}} (T_{\mathrm{H}} - T_{\mathrm{m}}),$$

and

$$Q_{\mathrm{D}} = m_{\mathrm{D}} C_{p_{\mathrm{D}}} (T_{\mathrm{m}} - T_{\mathrm{D}})$$

where Q = energy (kJ), m = mass (kg), C_p = specific heat capacity (kJ/kg/°C), T = temperature (°C), and subscripts H, m, and D refer to hydrolyzed material, mixture, and fluid used for dilution, respectively.

The energy lost by the hydrolyzed sludge is equivalent to that gained by the dilution water, i.e.

$$Q_{\mathrm{H}} = Q_{\mathrm{D}}$$

By substitution

$$m_{\mathrm{H}} C_{p\mathrm{H}} (T_{\mathrm{H}} - T_{\mathrm{m}}) = m_{\mathrm{D}} C_{p\mathrm{D}} (T_m - T_D)$$

rearranging for

$$T_{\mathrm{m}} = m_{\mathrm{H}}C_{p\mathrm{H}}T_{\mathrm{H}} + m_{\mathrm{D}}\,C_{p\mathrm{D}}T_{\mathrm{D}}/m_{\mathrm{H}}\,C_{p\mathrm{H}} + m_{\mathrm{D}}C_{pD} \tag{2.4}$$

For the worked example

m_{H} = mass of hydrolyzed sludge, which is the input sludge plus the steam:

$$= ((10{,}000\ \text{(t/year)} \times 1{,}000\ \text{(kg/t)})/16.5\%\ \text{(DS)}) + (950\ \text{(kg steam/t)} \times 10{,}000\ \text{(t/year)}$$

$$= 70{,}106{,}060\ \text{kg}$$

$C_{p\mathrm{H}} = 3.815$ kJ/kg/°C calculated by weighted average based on specific heat capacities for sludge and water, as the diluent is water, then $C_{p\mathrm{D}} = 4.2$ kJ/kg/°C assume $T_{\mathrm{H}} = 102$°C, m_D can be calculated by simple mass balance at 10% DS feed, mass to digester = ((10,000 (t/year) × 1,000 (kg/t))/10% (DS) = 100,000,000 kg therefore, the quantity of dilution water is the difference between this figure and the output from the hydrolysis unit:

$$m_{\mathrm{D}} = 100{,}000{,}000\ \text{kg} - 70{,}106{,}060\ \text{kg} = 29{,}839{,}940\ \text{kg}$$

Substituting into equation 2.4 for T_{m}

$$T_{\mathrm{m}} = (70{,}106{,}060\ \text{kg} \times 3.185\ \text{kg/kJ/°C} \times 102\text{°C} + 29{,}839{,}940\ \text{kg} \times 4.2\ \text{kJ/kg/°C} \times 25\text{°C})/70{,}106{,}060\ \text{kg} \times 3.185\ \text{kg/kJ/°C} + 29{,}839{,}940\ \text{kg} \times 4.2\ \text{kJ/kg/°C}$$

$$= 74\text{°C}$$

If the digester is fed at 38°C for example, then from heat transfer, the cooling requirement is

$$= 100{,}000{,}000\ \text{kg} \times 4.065\ \text{kg/kJ/°C}$$

(calculated by weighted average between hydrolyzed sludge and dilution water) (74°C – 38°C)

$$= 1.463 \times 10^{10}\ \text{kJ} = 4{,}065{,}000\ \text{kWhr/year or } 464\ \text{kW}$$

This is the minimum amount of cooling required, and the actual amount would have to include the efficiency of the cooling system.

2.1.4 Ways of reducing the energy demand of thermal hydrolysis further

Several configurations have been developed to reduce the parasitic energy demand of thermal hydrolysis, principle of which are those which treat only the waste-activated sludge fraction – as this is where most of the benefit lies – and treating digested sludge. While they both reduce or eliminate the requirement for

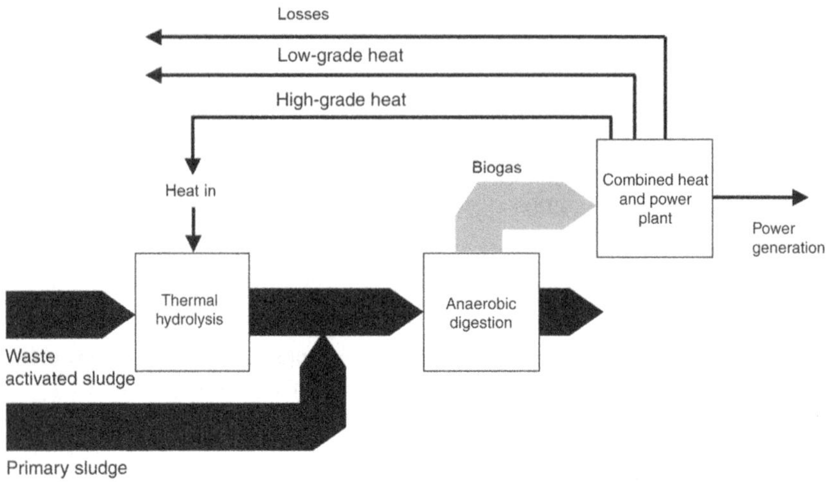

Figure 2.5 Thermal hydrolysis of only the waste-activated sludge fraction. Compare with Figure 2.3 for treatment of both primary and waste-activated sludge.

external energy, they have different influences on performance and impacts on sludge infrastructure.

2.1.4.1 Treatment of only waste-activated sludge

This configuration is suitable on sites which thicken primary and waste-activated sludge separately. In this instance, only the waste-activated sludge is processed, and the hot waste-activated sludge back mixed into the unhydrolyzed primary sludge. In this way the heat is not dumped, as it is when both primary and waste activated sludge are processed but used to heat the primary sludge prior to digestion. Although not all the sludge is processed, recent studies (Panter *et al.*, 2019) have shown that performance of digestion and dewatering are similar to systems where all sludge is processed. One hypothesis which has been suggested, is that this may be due to lower loading rates leading to lower concentrations of free ammonia in the digester. Figure 2.5 highlights the differences between hydrolyzing all sludge or only the waste-activated sludge portion, and is accompanied by Table 2.1.

As before, it is possible to quantify the difference in this configuration by means of a worked example.

Question. *How much energy is required for the previous example if only the waste-activated sludge fraction is hydrolyzed, and how much of the energy is available from high-grade heat recovered from co-generation? Assume volatile solids destruction of the primary sludge drops from 62% to 58% as it is not thermally hydrolyzed, and that the DS of the primary sludge is thickened to 5% DS.*

Table 2.1 Differences between treatment of all sludge and waste-activated sludge only

Thermal Hydrolysis of All Sludge	Thermal Hydrolysis of Only Waste-Activated Sludge
• Larger thermal hydrolysis plant • Parasitic energy demand • Needs dilution water • Needs cooler • Pathogen-free biosolids • Higher loading rate in digestion • Occasionally higher performance in downstream digestion and dewatering • Higher ammonia and refractory material production	• Smaller plant • No parasitic energy demand • Uses primary sludge as dilution water • Waste-activated sludge heats up primary sludge prior to digestion • Not pathogen-free biosolids • Typically, lower loading rate • Performance in digestion and dewatering similar to treatment of all sludge • Ammonia and refractory material are diluted by non-hydrolyzed sludge • Can be difficult to attain high DS in the thermal hydrolysis plant • High sludge viscosity can cause handling, mixing and heat-transfer problems

Answer.

First of all, calculate the energy demand of thermal hydrolysis:

The sludge is 40% waste activated, i.e. 10,000 t/year × 40% = 4,000 t/year = 4,000,000 kg/year

At 16.5% DS, wet flow = 4,000,000/16.5% = 24,242,424 kg wet weight of which 24,242,424 − 4,000,000 = 20,242,424 kg are water

Energy to heat sludge

$$= 4{,}000{,}000\,\text{kg} \times 1.5\,\text{kJ/kg/}^{\circ}\text{C} \times 75^{\circ}\text{C temperature increase}$$
$$= 450{,}000{,}000\ \text{kJ} = 125{,}000\,\text{kWhr}$$

Energy to heat water

$$= 20{,}242{,}424\,\text{kg} \times 4.2\,\text{kJ/kg/}^{\circ}\text{C} \times 75^{\circ}\text{C temperature increase}$$
$$= 6{,}376{,}363{,}560\ \text{kJ} = 1{,}771{,}212\ \text{kW hr}$$

Therefore, energy to heat sludge at 16.5% DS before efficiency losses

$$= 450{,}000{,}000 + 6{,}376{,}363{,}560 = 6{,}826{,}363{,}560\ \text{kJ}$$

Which after losses becomes 6,826,363,560/65% = 1.050×10^{10} kJ or 2,917,249 kWhr. This is equivalent to 292 kWhr/t DS based on 10,000 t dry sludge solids.

Based on the properties of steam at the required temperature and pressure, a quantity of 3,800,976 kg/steam are required.

Now calculate the energy available from the biogas, to determine the quantity of high-grade heat available.

In this instance, the biodegradability of the waste-activated sludge is the same as before, therefore, the biogas production from activated sludge is

$$= 1{,}200{,}000\ \mathrm{Nm^3\ biogas/year}$$

Now, the biodegradability of the primary sludge has dropped slightly to reflect that it has not been hydrolyzed. Therefore, biogas production from primary sludge is

$$\begin{aligned} &= 10{,}000\ (\mathrm{t/year}) \times 1{,}000\ (\mathrm{kg/t}) \times 60\%\ (\text{primary sludge}) \\ &\quad \times 75\%\ (\text{volatile fraction}) \times 58\%\ (\text{volatile solids destruction}) \\ &\quad \times 1.1\ (\mathrm{Nm^3/kg\ VS_{destroyed}}) \\ &= 2{,}871{,}000\ \mathrm{Nm^3\ biogas/year.} \end{aligned}$$

Therefore, total biogas production = 1,200,000 $\mathrm{Nm^3}$ biogas/year + 2,871,000 $\mathrm{Nm^3}$ biogas/year = 4,071,000 $\mathrm{Nm^3}$

Energy within gas = 4,071,000 × 65% (methane by volume) × 10.6 (kWhr/$\mathrm{m^3}$ methane) = 28,049,190 kWhr.

Therefore, prior to energy recovery the thermal hydrolysis plant needs 2,917,249/28,049,190 = 10.4% equivalence of the biogas energy.

How much of this energy can be recovered from high-grade heat?

The biogas energy goes into a co-generation unit where 25% is converted to high-grade heat and the engine is 92% available.

Therefore, the high-grade heat available = 28,049,190 kWhr × 25% (conversion to high-grade heat) × 92% (engine efficiency) = 6,451,314 kWhr.

As this number is greater than the demand of 2,917,249 kWhr, therefore, all the energy required can be provided by high-grade heat and there is no parasitic energy demand for the process, therefore, there is no requirement for auxiliary fuel in spite of a slight reduction in biogas energy production.

As mentioned, the hot sludge is typically mixed with the cool unheated primary sludge to transfer heat to it and therefore reduce the heating demand of the ensuing digestion plant.

Question. *What is the digester space heating requirement assuming no radiative and convective heat losses? What is the inlet DS and temperature entering downstream anaerobic digestion?*

In this instance, equation 2.4 can be used to determine temperature, but this time instead of clean water being the diluent, it is primary sludge.

Wet mass of primary sludge

= (10,000 (t DS/year) × 1,000 (kg/t) × 60% (primary sludge))/5% (DS)

= 120,000,000 kg

Wet mass of primary sludge (m_D)

= (10,000 (t DS/year) × 1,000 (kg/t) × 60% (primary sludge))/5% (DS)

= 120,000,000 kg

Wet mass of hydrolyzed sludge (m_H)

= (10,000 (t DS/year) × 1,000 (kg/t) × 40% (primary sludge))/16.5% (DS)

+ 3,800,976 (kg steam)

= 28,043,400 kg.

For the example

$C_{pH} = 3.815$ kJ/kg/°C, the specific heat capacity for the primary sludge can be calculated by weighted average using specific heat capacities for water and sludge as 4.2 and 1.5, respectively. Then $C_{pD} = 4.065$ kJ/kg/°C

assume $T_H = 102$°C, and temperature of primary sludge is 25°C = T_D

Substituting into equation 2.4 for T_m

$$T_m = (28{,}043{,}400\text{ kg} \times 3.815\text{ kg/kJ/°C} \times 102\text{°C} + 120{,}000{,}000\text{ kg} \times 4.065\text{ kJ/kg/°C} \times 25\text{°C})/28{,}043{,}400\text{ kg} \times 3.815\text{ kg/kJ/°C} + 120{,}000{,}000\text{ kg} \times 4.065\text{ kJ/kg/°C}$$

$$= 38.9\text{°C}$$

If the digester is fed at 38°C for example, then from heat transfer, the cooling requirement is

= 148,043,401 kg (sum of hydrolyzed flow and primary sludge) × 4.02 kg/kJ/°C (weighed average specific heat capacity of hydrolyzed flow and thickened primary sludge) (38.9 – 38°C)

= 148,043,401 kg (sum of hydrolyzed flow and primary sludge) × 4.02 kg/kJ/°C (weighed average specific heat capacity of hydrolyzed flow and thickened primary sludge) (38.9 – 38°C)

= 525,621,025 kJ = 148.783 kWhr/year or 17 kW

This quantity of cooling is negligible and realistically at full scale it is unlikely that a cooler would be installed at all. Coincidentally, the sludge composition used in this example (60:40 primary:activated sludge), occurs when cooling or heating requirements are at a minimum. At lower fractions of primary sludge there is a demand to cool the sludge down, while at higher fractions, there is a need to increase sludge temperature as shown in Figure 2.6.

Regarding the DS of the mixture, this is simply determined by weighed average between the flows of primary sludge 120,000,000 kg at 5% DS and 28,043,400 kg of hydrolyzed material at 14.26% DS

$$= ((120{,}000{,}000\ (\text{kg}) \times 5\%\ (\text{DS}) + 28{,}043{,}400\ (\text{kg}) \times 14.26\%\ (\text{DS}))/(120{,}000{,}000\ (\text{kg}) + 28{,}043{,}400\ (\text{kg}))$$
$$= 6.75\%\ \text{DS entering the digestion plant}$$

Note, there is an example of how processing only activated sludge impacts operating costs in Chapter 7. This configuration is popular when enhanced treated sludge is not required, or it precedes thermal processing such as drying.

Data from full-scale applications bearing this configuration shows greater than expected performance. Although, only the biological sludge is processed, the overall digestion and dewatering performance is almost similar to that experienced when all the sludge is processed. This provides anecdotal evidence that thermal hydrolysis is largely ineffective on primary sludge, but better performance may also be due to lower levels of ammonia toxicity (Panter *et al.*, 2019).

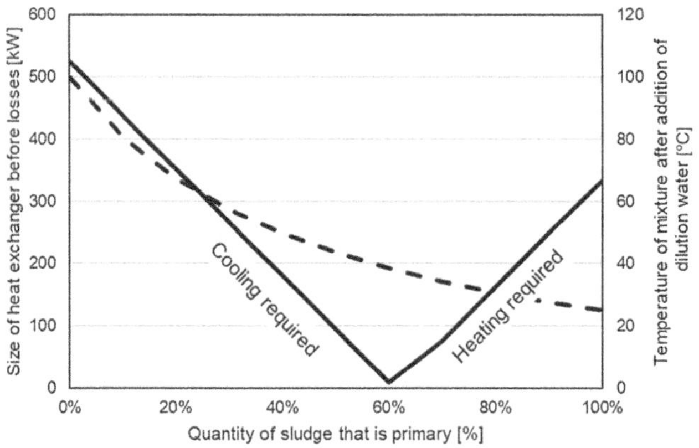

Figure 2.6 Influence of sludge composition on heat exchanger size (primary *y* axis, full-line) and temperature of mixture of hydrolyzed and non-hydrolyzed sludge (secondary *y* axis, dotted line) for thermal hydrolysis plant processing only waste-activated sludge fraction comprising 40% of 10,000 t DS per year. Digester fed at 38°C.

To improve the energy balance of Psyttaia WWTP in Greece, thermal hydrolysis was installed to treat a fraction of the biological sludge (Zikakis *et al.*, 2017). The facility has a drying plant downstream. After installation of pre-treatment, biogas yield increased from 390 to 450 Nm^3/kg TS fed to the digester. When combined with improved dewatering performance from 21/22% DS to over 31% DS, this reduced the energy demands of the dryer by approximately 40%, which freed up biogas for the production of renewable energy.

2.1.4.2 Treatment of digested sludge

In principle, digested sludge exiting a digester contains material which is harder to digest and therefore amenable to thermal hydrolysis. By processing only digested sludge, the thermal hydrolysis plant and concomitant energy demand is smaller. A trait this configuration shares with is the processing of only activated sludge described in the previous section. However, unlike processing only activated sludge, this configuration further benefits by treating all of the sludge, therefore, the treated sludge maintains enhanced treated status. This configuration can be achieved either on site, or off-site where dewatered digested biosolids are transported elsewhere for treatment. If the facility is designed to reach standard treatment in the first stage of digestion, it is possible to have a high level or redundancy and flexibility with this configuration. Figure 2.7 shows the basics of this configuration.

The following points distinguish this application of thermal hydrolysis to others:

- Smaller plant
- No parasitic energy demand
- Pathogen-free biosolids
- Needs more digestion capacity compared to treatment of all sludge
- Higher overall performance

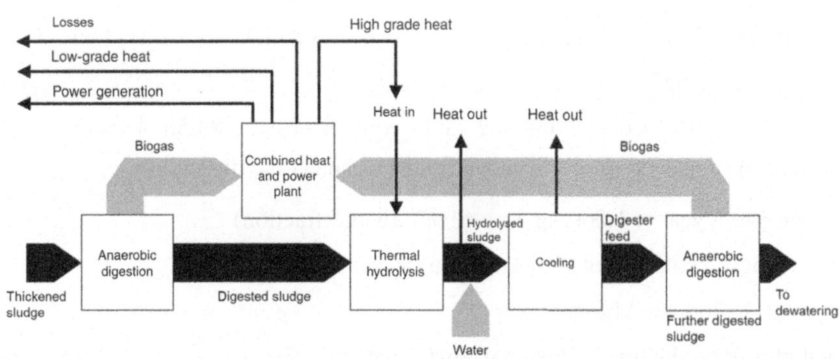

Figure 2.7 Thermal hydrolysis of digested sludge prior to further anaerobic digestion. Compare with Figure 2.3 for treatment of both primary and waste-activated sludge.

- Benefits from series digestion
- Can be on or off-site
- High level of redundancy
- Higher loading rate in second digestion stage due to lower volatile solids concentration

As before, it is possible to quantify the difference in this configuration by means of a worked example.

Question. *Assume that the raw sludge from the previous example is digested. The quantity of sludge remaining is 6,340 t DS with DS content of 3.2% and volatile solids content of 61%. The plant produced 3,711,000 Nm^3 biogas. It is proposed to thicken this sludge to 16.5% DS and then thermally hydrolyze it, assume that 50% of the remaining sludge is biodegradable, biogas yield is 1 Nm^3/t DS digested, and that after convection and radiative heat losses, the temperature of the digested sludge is 30°C, and dilution water is available at 25°C. Determine the energy required by the thermal hydrolysis system and the energy available from high-grade heat assuming biogas from both first and second stages of digestion is available? How large is the cooling heat exchanger?*

Answer.
As with the previous examples, start with calculation of the energy and steam demand, using equation 2.1:

Where

$$m_s = 6{,}340 \text{ t} \times 1{,}000 \text{ kg/t} = 6{,}340{,}000 \text{ kg}$$

$$m_w = ((6{,}340 \text{ t} \times 1{,}000 \text{ kg/t})/16.5\%) - (6{,}340 \text{ t} \times 1{,}000 \text{ kg/t})$$

$$= 32{,}084{,}242 \text{ kg}$$

$$\Delta T = 102 - 30°\text{C} = 72°\text{C}$$

Substituting:

$$= (6{,}340{,}000 \text{ kg} \times 1.5 \text{ kJ/kg/°C} \times 72°\text{C} + 32{,}084{,}242 \text{ kg} \times 4.2 \text{ kJ/kg/°C} \times 72°\text{C})/65\%$$

$$= 1.598 \times 10^{10} \text{ kJ} = 4{,}438{,}887 \text{ kWhr equivalent to } 5{,}783{,}566 \text{ kg steam}$$

Quantity of biogas produced in second stage of anaerobic digestion

$$= 6{,}340 \text{ t/year} \times 1{,}000 \text{ kg/t} \times 61\% \text{ (volatile fraction)} \times 50\% \text{ (volatile solids destruction)} \times 1.0 \text{ Nm}^3/\text{kg VS}_{\text{destroyed}}$$

$$= 1{,}933{,}700 \text{ Nm}^3 \text{ biogas/year}$$

Total biogas production inclusive of first stage:

$$= 3{,}711{,}000 \text{ Nm}^3 + 1{,}933{,}700 \text{ Nm}^3 = 5{,}644{,}700 \text{ Nm}^3$$

Energy within gas = 5,644,700 × 65% (methane by volume)

× 10.6 kWhr/m^3 methane

= 38,891,983 kWhr

Therefore, prior to energy recovery the thermal hydrolysis plant needs 4,438,887/38,891,983 = 11% equivalence of the biogas energy.

This increases to 20% biogas equivalents if only the biogas produced in the second stage of digestion is available.

The biogas energy goes into a co-generation unit where 25% is converted into high-grade heat and the engine is 92% available.

Therefore, the high-grade heat available = 38,891,983 kWhr × 25% (conversion to high-grade heat) × 92% (engine efficiency) = 8,945,156 kWhr.

As this figure is greater than the requirement of 4,438,887 kWhr, there is no requirement for external energy, as is the case when only waste-activated sludge is processed. Incidentally, the energy required for thermal hydrolysis is also less than the waste heat produced when only considering the biogas production from the second stage of digestion, making this configuration totally independent of external energy sources.

With respect to the cooling requirements, it is first necessary to determine the influence of addition of dilution water to reduce loading rate into the second stage of digestion. Assuming the second digestion stage is fed at 10% DS as with the previous examples:

at 10% DS feed,mass to digester = ((6,340(t/year) × 1,000(kg/t))/10%(DS)

= 63,400,000 kg.

Therefore, the quantity of dilution water is the difference between this figure and the output from the hydrolysis unit (32,084,242 + 3,513,978 = 35,598,220 kg):

For equation 2.4, m_D = 63,400,000 kg − 35,598,220 kg = 27,801,780 kg

Substituting this into equation 2.4 for T_m

T_m = (35,598,220 kg × 3.185 kg/kJ/°C × 102°C + 27,801,780 kg

× 4.2 kJ/kg/°C × 25°C)/35,598,220 kg × 3.185 kg/kJ/°C

+ 27,801,780 kg × 4.2 kJ/kg/°C = 63°C

If the digester is fed at 38°C for example, then from heat transfer, the cooling requirement is

= 63,400,000 kg × 4.065 kg/kJ/°C (calculated by weighed average

between hydrolyzed sludge and dilution water) (63 − 38°C)

= 6.443 × 10^9kJ = 1,789,729 kWhr/year or 204 kW

The example shows that digesting sludge prior to thermal hydrolysis not only makes the plant energy self-sufficient but also significantly reduces the size of heat exchanger required.

There are a few installations worldwide which are based on this configuration. Notably, see Billund (Chapter 6), Marquette-Lez-Lille, and Ljubljana. Additionally, Thames Water have been undertaking research into quantifying the impacts of this configuration compared with other formations of thermal hydrolysis on both, site and strategic level. In a recent costing exercise (Rus, 2018), showed that operating costs for standard digestion – once benefits had been accounted for – were approximately £86/t DS processed. These reduced to a cost of £29/t DS processed when thermal hydrolysis was introduced, and a revenue of £36/t DS for thermal hydrolysis of digested sludge. This revenue was due to a combination of enhanced biogas production and improved dewaterability when compared to standard pre-digestion thermal hydrolysis.

2.1.4.3 Use of thermal hydrolysis as a dewatering aid with liquor recycle

Looking through the literature reveals that the initial aim of thermal hydrolysis was not to improve the performance of anaerobic digestion, but rather to improve dewatering. Work between the 1950s and mid-1970s found that when raw undigested sludge was exposed to conditions of thermal hydrolysis, the resultant material dewatered routinely above 50% DS, and depending on sludge type, as high as 60% DS. Subsequent work has shown that, by reintroducing materials known to influence dewatering, downstream anaerobic digestion deteriorates the dewatering potential of the biosolids. Although the use of thermal hydrolysis with digestion improves dewatering by approximately 10% points compared with a case with no pre-treatment, the dewaterability would have been higher without the digestion process at all. With this configuration, thermal hydrolysis is installed downstream of the digestion plant immediately prior to dewatering. In this instance, the digested sludge is dewatered hot, and the centrate, now laden with biodegradable chemical oxygen demand (COD), is recycled to the digester inlet and digested as shown in Figure 2.8. At the time of writing, there is one full-scale plant in Europe and a pilot plant in China based on this configuration.

Studies from the full-scale facility in Europe show higher levels of volatile solids destruction and better dewaterability, when compared to other configurations. However, as the warm liquor stream is returned, the hydraulic retention time of the digestion plant is reduced. If this lowers too much, this configuration may not be viable, or it may be necessary to provide a purpose built high-rate anaerobic digester for the return flow (Kjorlaug *et al.*, 2015). Although performance of both digestion and dewatering are improved, the dewatered cake does not meet the requirements for advanced treatment, as defined in the US EPA 503 regulations.

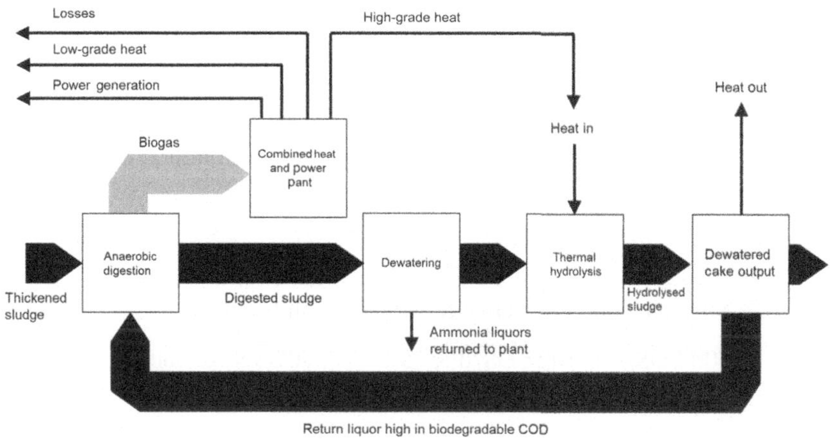

Figure 2.8 Thermal hydrolysis of digested sludge prior to further anaerobic digestion. Compare with Figure 2.3 for treatment of both primary and waste-activated sludge.

This is due to the pasteurization stage (thermal hydrolysis) being downstream of the vector attraction stage (anaerobic digestion). Subsequently, to meet the requirements of advanced treatment, the cake would have to be processed further, typically by drying or even by the addition of lime.

In the following example, one can see the impacts of this configuration.

Question. *A treatment plant digests 100 t DS/d of sludge. The sludge contains 75% volatile solids and has a COD:VS ratio of 1.6. The design loading rate is 2.5 kg VS/m^3/d based on sludge being fed at 5% DS. 50% of the COD is destroyed. Assume volatile solids destruction if 45%. After the upgrade, volatile solids destruction increases to 70%. Ambient air is at 14°C, and the digester is operated at 36°C. Convection and radiation losses in the digestion plant are approximately 30% of the space heating requirements. Assume the specific heat capacity of dried solids is 1.5 kJ/kg/°C. Initially, DS from dewatering was 22% DS with 95% capture rate. With the upgrade the output of dewatering increases to 47% DS with the same capture rate. The DS exiting thermal hydrolysis is 14% due to the addition of steam. The liquors returned to the front of the digestion plant are at 80°C. 60% of the COD in the return liquors is solubilized and the COD is 70% biodegradable. The polymer dose for dewatering is 8 kg/t DS and is made up in a 0.2% solution. Determine the following:*

(a) *The change in hydraulic retention time in the digestion plant?*
(b) *The increase in biogas production after the installation of thermal hydrolysis?*
(c) *The decrease in biosolids production after the upgrade?*
(d) *The reduction in the digestion plants heating demand*

Answer.

(a) To calculate the change in hydraulic retention time, it is necessary to determine the quantity of water returning in the recycle flow. This quantity also includes the polymer water. Therefore, it is necessary to find out the quantity of polymer required, which in turn is dependent on the solids exiting the digester.

Solids exiting the digester:

This is determined by mass balance on volatile solids.

Solids in-solids converted to biogas = solids out

$$100 \text{ t DS/d} - 100 \text{ t DS/d} \times 75\%\text{VS} \times 70\%\text{VS destruction}$$

$$= 100 \text{ tVS} - 52.5 \text{ tVS} = 47.5 \text{ t DS out}$$

Polymer dose:

$$47.5 \text{ t DS/d} \times 8 \text{ kg/t DS} = 380 \text{ kg polymer/d} = 0.38 \text{ t/d}$$

This is made up from an 0.2% solution, therefore quantity of water required to mix polymer

$$= 0.38/0.2\%$$

$$= 190 \text{ m}^3$$

For the dewatering stage, 14% DS (exiting thermal hydrolysis) are thickened to 47% DS with a 95% capture rate.

Therefore, cake quantity

$$= 47.5 \text{ t DS/d} \times 95\% \text{ capture}/47\% \text{ DS})$$

$$= 96 \text{ m}^3/\text{d}$$

Therefore, there is approximately one wet tonne biosolids produced per dry tonne of raw sludge digested. This is typical of full-scale performance.

Flow entering dewatering:

$$= 47.5 \text{ t DS/d}/14\%$$

$$= 339 \text{ m}^3$$

Flow returning to the digester is the difference between inlet and outlet flow from dewatering plus the polymer water.

$$= 339 \text{ m}^3/\text{d} - 96 \text{ m}^3/\text{d} + 190 \text{ m}^3/\text{d}$$

$$= 433 \text{ m}^3/\text{d}$$

Loading rate to digester is 2.5 kg $\text{VS/m}^3/\text{d}$

Therefore, digester volume is

$$= 100 \text{ t DS/d} \times 75\%\text{VS} \times 1{,}000\text{kg/t}/2.5 \text{ kg VS}$$

$$= 30{,}000 \text{ m}^3/\text{d}$$

The digester is being fed at 5% DS

$$= 100/5\% = 2{,}000 \text{ m}^3/\text{d}$$

Therefore, retention time in digester prior to upgrade

$$= 30{,}000 \text{ m}^3/\text{d}/2{,}000 \text{ m}^3/\text{d}$$
$$= 15 \text{ d}$$

Now 433 m^3/d are being returned from dewatering, so new retention time is

$$= 30{,}000 \text{ m}^3/\text{d}/(2{,}000 \text{ m}^3/\text{d} + 433 \text{ m}^3/\text{d})$$
$$= 12.3 \text{ d}$$

Therefore, the upgrade has resulted in a drop in retention time of ca 2.5 d (<20%).

(b) Increase in biogas production. This can be determined by simple balance on COD.

COD entering digester

$$= 100 \text{ t DS/d} \times 75\% \text{ VS} \times 1.6 \text{ COD:VS ratio} \times 1{,}000 \text{ kg/t}$$
$$= 120{,}000 \text{ kg/d}$$

COD destroyed before upgrade

$$= 120{,}000 \text{ kg/d} \times 50\% \text{ COD destroyed}$$
$$= 60{,}000 \text{ kg/d}$$

From stoichiometry, 1 kg COD destroyed produces 0.35 m^3 methane under standard temperature and pressure. Therefore, methane produced from the COD

$$= 60{,}000 \text{ kg/d} \times 0.35 \text{ m}^3 \text{ methane/kg COD}$$

$$= 21{,}000 \text{ m}^3\text{methane, or } 32{,}308 \text{ m}^3 \text{ biogas with 65\% methane within it.}$$

(For comparison, the gas production based on destruction of volatile solids and a gas yield of 0.9 m^3/kg VS destroyed = 30,375 m^3 biogas.)

After upgrade, COD is being returned from the dewatering stage. From earlier, 60,000 kg/d COD exit the digester. From the assumptions, this COD is 60% solubilized, of which 70% is biodegradable.

Therefore, the biogas production from the return stream

$= 60{,}000\ \text{kg/d} \times 60\%\ \text{solubilized} \times 70\%\ \text{biodegradable}$

$\times\ 0.35\ \text{m}^3\ \text{methane/kg COD destroyed}$

$= 8{,}820\ \text{m}^3/\text{methane}$

$= 13{,}569\ \text{m}^3/\text{biogas}$ assuming similar composition to the earlier.

Therefore, after the upgrade, the biogas production has increased by 13,569 m^3/biogas to 45,877 m^3 biogas/d. An increase of >40%. In total, >70% of the COD has now been destroyed.

In this example, each cubic meter of liquor returned yields 22 cubic meters of biogas.

(c) Decrease in biosolids production after upgrade. Some of this answer has already been determined earlier.

From before, cake quantity after upgrade

= 96 wet t/d

Prior to the upgrade, volatile solids destruction was 45% and dewatering was at 22% DS with a 95% capture rate.

DS exiting the digester prior to upgrade:

= 100 t DS/d – 100 t DS/d × 75% VS × 45% VS destroyed
= 66.25 t DS/d

Therefore, cake production

= 66.25 t DS/d × 95% capture/22% DS
= 286 wet t/d

This is a reduction of 190 wet t/d (~65% compared to baseline prior to upgrade)

(d) Influence on digester heat balance.

Space heating prior to upgrade is simply determined from basic thermodynamics as described earlier in this chapter.

100 t DS/d contain 100 × 1,000 kg = 100,000 kg dry material

At 5% DS, the wet weight of sludge is 100,000/5% = 2,000,000 kg of which 100,000 kg are dry and (2,000,000 – 100,000=) 1,900,000 kg are wet.

Energy to heat sludge

$= 100{,}000\ \text{kg} \times 1.5\ \text{kJ/kg/°C} \times (36 - 14\text{°C} = 22\text{°C})$
temperature increase $= 3{,}300{,}000\ \text{kJ}$

Energy to heat water

$= 1{,}900{,}000\ \text{kg} \times 4.2\ \text{kJ/kg/°C} \times 22\text{°C}$ temperature increase

$= 175{,}560{,}000\ \text{kJ}$

In total, this is 178,860,000 kJ = 49,683 kWhr.

From the assumptions, 30% additional heat is lost, therefore, the heat demand of the digester is

$= 49{,}683\ \text{kWhr} + 49{,}683\ \text{kWhr} \times 30\%$

$= 64{,}588\ \text{kWhr}$

After upgrade, 433 m^3/d are being returned at 80°C and blended with the original flow of 2,000 m^3/d which is at ambient temperature of 14°C.

Using equation 2.4 from earlier:

$$T_m = m_H C_{pH} T_H + m_D C_{pD} T_D / m_H C_{pH} + m_D C_{pD}$$

$= (433{,}000$ kg water* $\times 4.2\ \text{kJ/kg/°C} \times 80\text{°C} + 2{,}000{,}000$* kg input flow $\times 4.065\ \text{kJ/kg/°C} \times 14\text{°C})/433{,}000$ kg water* $\times 4.2\ \text{kJ/kg/°C} + 2{,}000{,}000$* kg input flow $\times 4.065$** $\text{kJ/kg/°C}) = 26\text{°C}$

*Assumes the sludge to have the same density as water.
**Weighed average specific heat capacity.

To determine the new heating requirements, use the same equation to determine space heating with 26°C rather than 14°C as the lower temperature, that is the sludge needs to increase in temperature by 10°C.

Energy to heat sludge

$= 100{,}000\ \text{kg} \times 1.5\ \text{kJ/kg/°C} \times (10\text{°C})$ temperature increase

$= 1{,}500{,}000\ \text{kJ}$

Energy to heat water

$= 1{,}900{,}000\ \text{kg} \times 4.2\ \text{kJ/kg/°C} \times 10\text{°C}$ temperature increase

$= 79{,}800{,}000\ \text{kJ}$

In total, this is 81,300,000 kJ = 22,583 kWhr.

From the assumptions, 30% additional heat is lost, therefore, the heat demand of the digester is

$$= 22{,}583\ \text{kWhr} + 22{,}583\ \text{kWhr} \times 30\%$$
$$= 29{,}358\ \text{kWhr}$$

Therefore, the heating requirements of the digester have reduced by 64,588 − 29,358 kWhr = 35,230 kWhr (a reduction of 54%).

The example is loosely based on performance from a full-scale plant in Geiselbullach Water Resource Recovery Facility, near Munich, Germany.

Geiselbullach Water Resource Recovery Facility produces approximately 4,400 t sludge DS annually with a high volatile solids content of 83%. The sludge composition is typical of a European plant, comprising 60% primary with 40% waste-activated sludge, based on biological phosphorous removal. The facility has mesophilic anaerobic digestion with retention time of 22 d and managed a volatile solids reduction of approximately 50%, prior to installation of the WAS-only thermal hydrolysis plant.

Deployment of thermal hydrolysis on only the biological sludge saw an increase in volatile solids destruction to 65%, and this increased further to 71% with downstream application. Accordingly, gas production increased from 472 to 596 $Nm^3/t\ DS_{fed}$. In conjunction with improved dewatering by 18 percentage

Table 2.2 Comparison between pre- and post-digestion thermal hydrolysis

Parameter	Pre-digestion Thermal Hydrolysis	Post-digestion Thermal Hydrolysis
Size of plant required	100%	60%
Steam demand $[kg_{steam}/kg_{DS\ sludge}]$	0.8–1.0	0.5–0.6 or higher*
Volatile solids destruction	50–65%	70%+
Dewatering improvement**	+10 points DS	> +20 points DS
Typical cake production (wet t/dry t digested)	1.7	<1
Cooler required	Yes	No
Digestion capacity required**	30–40%	ca 100%
Enhanced treated	Yes	No***

*Steam demand depends on operating temperature in this instance (see earlier).

**Compared to standard anaerobic digestion.

***Potentially can comply with enhanced treated biosolids as defined by US EPA 503 regulations if cake is greater than 70% DS.

Table 2.3 Performance differences between different configurations of thermal hydrolysis

Configuration	Biogas Production (m^3/t DS_{fed})	Biogas (m^3_{gas}/m^3_{dig})	Cake Production (Wet t/Dry t_{proc})	Steam Energy Required (% of Biogas)	Steam Energy Available (% of Required)	Parasitic Demand (% of Biogas)
Standard digestion	320	0.8	2.70	N/A	N/A	N/A
Before digestion	415	3.0	1.65	25–30%	>90%	<5%
Only biological sludge	390	2.8	1.80	10%	100%	0
Digested sludge	505	2.1	1.25	15%	100%	0
As dewatering aid	555	4.0	0.95	<10%	100%	0

points, this reduced biosolids cake production by 60%. Additionally, heating requirements for the digester reduced from 5,540 to 2,800 kWhr (Barber *et al.*, 2017; Kjorlaug *et al.*, 2015; Kolovos *et al.*, 2016).

In other interesting developments, this configuration has been used with higher thermal hydrolysis operating temperatures and different dewatering devices to produce further improvements in dewatering (Choo-Kun *et al.*, 2018). In their study, Choo-Kun and co-workers tested thermal hydrolysis operating temperatures between 150°C and 250°C at retention times between 0.5 and 2.5 h. The material was then digested under standard mesophilic conditions and then dewatered at temperatures between 20°C and 75°C. These tests were then validated at pilot-scale on a plant of 24,000 population equivalents in China. The work showed that dewaterability could improve to approximately 65% DS corresponding to a cake reduction of 70% compared to the baseline scenario. Lower heating value of the material was measured at 9 MJ/kg_{cake} which was comparable to municipal waste. Biogas production

Table 2.4 Additional differences between thermal hydrolysis configurations

Configuration	Good For	Challenges
Thermal hydrolysis of all sludge prior to digestion	• Increasing digestion capacity • Enhanced treated biosolids	• Inherent energy demand • Larger plant • Higher steam demand
Thermal hydrolysis of only biological sludge	• Reducing energy demand • Decreasing capital costs • Eliminating need for cooler	• Not enhanced treated sludge
Thermal hydrolysis of digested sludge	• Reducing energy demand • Decreasing capital costs • Enhanced treated biosolids • Inherent flexibility (first digestion stage can be designed to meet standards for standard treated biosolids)	• Needs sufficient digestion capacity
As a dewatering aid	• Significant reduction in biosolids (good for high reuse costs) • Enhanced dewatering (good for downstream thermal processing) • Inherent flexibility (can be switched off without adversely affecting outputs)	• Decreases digestion capacity • May not be enhanced treated • Potential issues with effluent quality • Dewatering of hot biosolids

increased by over 30%, and in spite of higher thermal hydrolysis operating temperatures, this almost made the facility self-sufficient for energy. Table 2.2 shows the differences of this configuration when compared with pre-digestion thermal hydrolysis.

2.1.5 Summary of different configurations

As shown in the previous sections, there are several ways to employ thermal hydrolysis. The value of these depends critically on the specific requirements of a particular project (Solley & Chhillar, 2019) and differences are shown in Table 2.3.

In addition to the above, additional factors influence choice. A summary is given in Table 2.4.

REFERENCES

Barber B., Nilsen P. J. and Christy P. (2017). Cambi SolidStream®: Thermal hydrolysis as a pre-treatment for dewatering to further reduce operating costs. *Proceedings of the Water Environment Federation*, **2017**(5), 5070–5083.

Choo-Kun M., Prevot C., Pardo P. E., Bourdais J. L., Judenne E., Camacho P., Lebars F. and Fournot McGill A. (2018). Moving towards Maximum Biosolids Reduction: Ultra-dewatering of Sludge. Proceedings of IWA Singapore International Water Week conference, Singapore.

Exelys company literature. http://technomaps.veoliawatertechnologies.com/processes/lib/municipal/3472-EN_Brochure_Exelys_0516.pdf

Fernández-Polanco D. and Tatsumi H. (2016). Optimum energy integration of thermal hydrolysis through pinch analysis. *Renewable Energy*, **96**, 1093–1102.

Green D. and Southard M. Z. (2019). 9th edn, Perry's Chemical Engineering Handbook. McGraw-Hill Publishing.

Kjorlaug O., Nilsen P. J., Solheim O. E., Traa L. P. and Kruchen H. (2015). Cambi Solid-Stream® High Dry Solids Technology Development of Test Procedures and Report from the First Full-scale Installation in Germany.

Kolovos A., Kjorlaug O. and Nilsen P. J. (2016). Development and Assessment of a Model for Cambi's SolidStream Process Using "SUMO" WWTP Simulation Software. Proceedings of the 21st European Biosolids and Organic Resources Conference, 15–16 November, Edinburgh, Scotland.

Merry J. and Oliver B. (2015). A comparison of real AD plant performance: Howdon, Bran Sands, Cardiff and Afan, In Proceedings of 19th European Biosolids & Organic Resources Conference & Exhibition, Aquaenviro, UK.

Lancaster R. (2015). Thermal Hydrolysis at Davyhulme WWTW, One Year On. *Water Environment Federation (WEF) Residuals and Biosolids.*

Panter K., Fountain P. and Shana A. (2019). The Effects of Thermal Hydrolysis, Hydraulic Retention Time and Ammonia Concentration on Digestion Rates and Dewatering Across Digestion Sites in Thames Water, UK. Proceedings of WEF Residuals and Biosolids Conference, Fort Lauderdale, FL, pp. 1175–1185.

Pook M., Mills N., Heitmann M., Panter K. and Walley P. (2013). Exploring the upper limits of thermal hydrolysis at Chertsey STW. In Proceedings of 17th European Biosolids and Organic Resources Conference and Exhibition, Aquaenviro-CIWEM, UK.

Rus E. (2018). Thames Water and THP. Presentation given at Cambi Academy, Phoenix, Arizona.

Rus E. and Esparza L. (2017). Second Generation THP:ITHP. Proceedings of SludgeTech Conference, 9–13th July, London.

Rus E., Perrault A., Mills N., Shana A., Molokwu O. and Nilsen P. J. (2017). Optimizing THP – The Intermediate Thermal Hydrolysis Process. Proceedings of IWA Specialist Conference on Sludge Management, SludgeTech.

Solley D. and Chhillar R. (2019). Which Thermal Hydrolysis Process Configuration Do I Choose? In Proceedings of WEF/IWA Residuals and Biosolids Conference, Fort Lauderdale, FL.

Xu C. C. and Lancaster J. (2009). Treatment of secondary sludge for energy recovery. *Energy recovery*, Nova Science Publishers Incorporation, New York, NY, USA, 187–211.

Zikakis D., Chauzy J., Droubogianni I. and Georgakopoulos A. (2017). The Application of WAS-Only Thermal Hydrolysis at Psyttalia WWTW. Proceedings of IWA Specialist Conference on Sludge Management, SludgeTech, London.

Chapter 3

Impacts of thermal hydrolysis

3.1 INTRODUCTION

This chapter deals with the most significant impacts of thermal hydrolysis in more detail. That is to say, its influence on rheological properties of sludge, dewatering, production of recalcitrant compounds and concerns surrounding ammonia toxicity in digestion due to higher loading rates. Treatment of emerging contaminants is also referred to. The influence of thermal hydrolysis on biogas production is presented in Chapter 5, along with other benefits. This chapter also summarizes the various studies which have been conducted on understanding the influence of thermal hydrolysis on microbial communities and their evolvement during the digestion process.

3.2 INFLUENCE ON SLUDGE RHEOLOGY

Rheology change due to thermal hydrolysis is arguably the most important consequence of the technology on sludge treatment as it allows higher digester loading rates – due to increased ease of downstream transport – and aids in dewatering (Stickland, 2015). In the informative review by Eshtiaghi and co-workers (2013), sludge is described as a non-Newtonian shear thinning thixotropic fluid, which behaves as a thixotropic colloidal suspension at high shear rates but exhibits polymeric behavior at low shear. The flow behavior of waste-activated sludge has been successfully modelled using the

Author: William Barber
doi: 10.2166/9781789060287_0057

Herschel–Bulkley equation, namely

$$\sigma = \sigma_c + k\gamma^n \tag{3.1}$$

where σ is the shear stress (Pa); σ_c is the yield stress (Pa); k is the consistency index (Pa s^n) and n is the flow index. These variables are fundamentally influenced by temperature. More recently, studies have been looking closer at the rheological behavior of sludge which has been thermally hydrolyzed (Hii *et al.*, 2107, 2019; Zhang *et al.*, 2017).

Increasing temperature by thermal hydrolysis reduces viscosity in several ways. First, free water within the sludge fraction decreases in viscosity in accordance with Arrhenius" law in a reversible fashion (Al-Shemmeri, 2012; Eshtiaghi *et al.*, 2013). In addition, material may be thermally destroyed within the sludge in a partly reversible way, for example the denaturation of protein or destruction of extracellular polymers (Farno *et al.*, 2015). Finally, interactions between compounds released due to heat treatment have been found to influence rheology (Forster, 1983). Conversely, the decrease in particle size due to steam explosion (Neyens *et al.*, 2004) is expected to increase viscosity by increasing the surface area of particles enhancing interactions (Pevere *et al.*, 2006). However, looking at rheology data from thermal hydrolysis, this increase is clearly outweighed by the aforementioned factors. Lotito and Lotito (2014) elegantly summarize the parameters influencing sludge rheology to be: sludge type; density; solids content; particle size and distribution; settleability; abrasiveness; particle friability; surface charge; liquid phase conductivity; pH and surface chemistry (Forster, 1983) amongst others.

A number of studies have shown that increasing the thermal hydrolysis operating temperature decreases apparent viscosity between temperatures of approximately 130–180°C (Higgins *et al.*, 2015; Urrea *et al.*, 2015). The decrease in viscosity due to temperature difference alone would result in a 30% drop using Arrhenius" relationship, but this is insufficient to describe the relative reductions observed from analysis of various workers data (Bougrier *et al.*, 2006; Dawson & Ozgencil, 2009). Additionally, when sludge is cooled down, irreversibility is noted as commented on by Baudez *et al.* (2013), who found that yield stress of sludge exposed to heat treatment and then cooled to its initial temperature was lower than in an un-exposed aliquot. Farno *et al.* (2014) also reported similar results on sludge treated at temperatures between 20°C and 80°C, which were returned to starting temperatures. Liao's group (2016) saw a drop in viscosity of 48.6% due to heating to 80°C, and when cooled to 33°C sludge viscosity remained a third lower than an unheated sample.

In an interesting study by Hii and co-workers (2017) sludge at 7% dry solids (DSs) was exposed to temperatures between 80°C and 145°C, for reaction times up to 1 hr. Measurements of high shear (at 600/s) viscosity and also yield stress were measured in the untreated sample, throughout the thermal hydrolysis process and also in the sample when it was cooled back down to ambient conditions. The authors found that once cooled down, the high shear viscosity and yield stress remained lower than in the untreated sample by 28%, 38%, <53%, 70% and 89% for temperatures of 80°C, 100°C, 120°C, 130°C and 145°C, respectively. The workers found that the majority of the reduction in viscosity and yield stress occurred within 10 min, and that the majority of solubilization of COD within 20 min of heat treatment. This has implications on the retention time required for thermal hydrolysis.

The phenomenon of rheological properties remaining lower after cooling down preceded by heat treatment has been termed "thermal history." In the study of Farno *et al.* (2014) the measured soluble COD of sludge which was heated and cooled down was significantly higher than the soluble COD of initial sludge such that a correlation was found between an increase in soluble COD and a decrease in yield stress. Similar relationships have been found with yield stress and high shear viscosity (Hii *et al.*, 2017). The difference could potentially be attributed to irreversible thermal denaturing of protein (Anson, 1954), which influences viscosity by unfolding and increasing quantities of water bound by protein molecules. By increasing the hydrodynamic radius of the molecules, viscosity is increased (Anson & Mirsky, 1932 cited in Farno *et al.*, 2014). The denaturing of proteins in this way, along with the rate and order of solubilization of material during thermal hydrolysis may partially explain the reduction in viscosity with increasing reaction temperatures. Bougrier's research team (2006), and subsequently Hii *et al.* (2017) demonstrated that sludge became more Newtonian in behavior as reaction temperature increased, noted by an increase in the dimensionless flow behavior index n when using the power law. This was also observed by Urrea's team (2015) looking at thermal hydrolysis of activated sludge. Figure 3.1 is a plot based on work presented by Bougrier's group (2006) using the Ostwald–de Waele relationship as follows:

$$\tau = 4.7 \times 10^{-3}\gamma^{0.89} \tag{3.2}$$

where τ = yield stress (Pa); γ = shear rate (1/s) and apparent viscosity η is τ/γ. The interested reader is referred to the manuscript by Bougrier *et al.* (2006).

The graph clearly demonstrates the viscosity drops as a consequence of thermal hydrolysis and suggests that the Ostwald–de Waele relationship is valid in the range shown. However, as highlighted in the excellent review by Eshtiaghi *et al.* (2013) the use of this relationship fails at predicting non-Newtonian fluid behavior at very high shear rates where the viscosity of the fluid remains higher than water. Indeed, the review mentions that at both high and low shear rates different analytical equipment are necessary to make rheology measurements to overcome limitations in testing protocol due to wall-slip or end effects. Hii's team (2017) has provided a comprehensive compilation of variables with which it is possible to model changes in rheological behavior due to heat treatment using similar relationships to those described here.

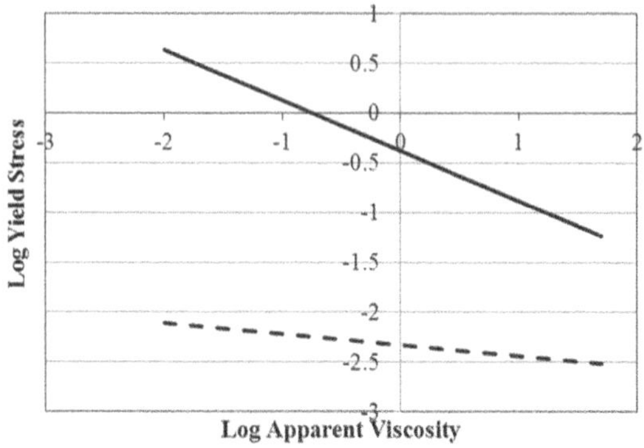

Figure 3.1 Influence of thermal pre-treatment (170°C) on sludge (at 20 g/l TS with 76% VS) on apparent viscosity. *Key*: Untreated sludge (full line); thermally hydrolyzed (dashed line). Plot using data presented in Bougrier *et al.* (2006).

When thermally hydrolyzed sludge is prepared for digestion, its viscosity will increase by a factor of at least 3 in accordance with Arrhenius" Law, although the addition of dilution water will influence this further. The analysis of unpublished data shows an increase in viscosity of between 2 and 6 times depending on the initial reaction temperature. However, the viscosity still remains lower than that of sludge unexposed to thermal conditioning, as shown by Hii *et al.* (2017). The change in viscosity post thermal hydrolysis, due to heat exchange and also addition of cooling water is of critical importance when designing sludge processing equipment (Hii *et al.*, 2019).

It is well understood that the rheological properties of primary and activated sludge are profoundly different and require different relationships to model their behavior (Eshtiaghi *et al.*, 2013; Markis *et al.*, 2014), again, only recently is this variable being studied. In a study looking at optimizing sludge pumping and mixing systems, Markis' group (2014) suggested that primary sludge behaved like a colloidal suspension with flocs governed by weak van der Waals forces, while activated sludge rheology was influenced by extracellular polymers held together by hydrogen and electrostatic forces characteristic of a gel–matrix. It is considered that 80% of the mass of activated sludge is comprised of extracellular material which influences viscosity (Neyens *et al.*, 2004). This results in the viscosity of activated sludge being much higher than that of primary (Dawson & Sotiriadis, 2007; Lotito & Lotito, 2014). Markis *et al.* (2016) demonstrated that they could increase the viscosity of a sludge mixture by increasing the proportion of activated sludge within it. These factors become progressively important for thermal hydrolysis systems processing only activated or high levels of activated sludge as is becoming increasingly vogue.

With specific attention to activated sludge, Hii's team (2019) measured *in-situ* rheological parameters (high shear viscosity, consistency index and yield stress) to characterize flow behavior of sludge at concentrations between 7% and 13% DSs at temperatures up to 130°C, with respect to developing predictive tools. These tools are useful for understanding sludge behavior after thermal hydrolysis, during cooling and pumping into anaerobic digestion. Thermal hydrolysis significantly altered the rheological properties of activated sludge with apparent viscosity and yield stress of samples – which had been processed and then cooled back to 25°C – remaining 14–72% and 9–60% lower respectively depending on treatment temperature between 80°C and 140°C (Hii *et al.*, 2019). The authors showed that the impact of thermal hydrolysis treatment temperature, duration of treatment and sludge concentration impacted high shear viscosity, consistency index and yield stress in linear, logarithmic and power-law relationships, respectively. Thermally hydrolyzed activated sludge exhibited a gel-like viscoelastic behavior; however, storage and loss moduli were reduced fundamentally and remained lower after cooling at 8–39% and 13–50% of the results obtained for untreated samples accordingly depending on the reaction temperature. These results showed a weakening of the sludge's elastic characteristics. These characteristics were also observed by Zhang *et al.* (2017).

In an informative study, Lotito and Lotito (2014) showed that the viscosity of digested sludge was higher than that of raw undigested sludge, and that both shear rates and shear stresses increase with DS content (Jiang *et al.*, 2014). This may be due to the generation of soluble microbial products during digestion (Barker & Stuckey, 1999), which ultimately influence dewaterability characteristics of the digestate (Christensen *et al.*, 2015; Lu *et al.*, 2015; Neyens & Baeyens, 2003; Neyens *et al.*, 2004).

Zhang and associates (2017) looked at using rheological measurements to determine the performance of thermal hydrolysis at two temperature regimes of 60–90°C and 120–180°C. The authors looked at 14% and 18% DS dewatered sludge. They found that under the lower temperature regime, time of heating was most important with respect to solubilization (of COD, proteins and polysaccharides) and changes in viscoelastic behavior. However, temperature was the dominant factor in the higher temperature range. For the lower temperature regime, the ratio of treated storage modulus (G') to untreated modulus (G'_0) showed a strong linear relationship with a reduction in solubility corresponding to increasing (G'/G'_0). This relationship changed to a logarithmic one (of the form solubility $= a - b\ln(G'/G'_0)$) in the higher temperature range (Zhang *et al.*, 2017).

3.3 INFLUENCE ON DEWATERING

There is unanimous agreement in the literature relating to the positive influence thermal hydrolysis has on sludge dewaterability ever since the earliest work when it was viewed as a dewatering aid (Everett, 1972; 1974; Lumb, 1940, 1951; Sheerwood & Phillips, 1970) and later to improve anaerobic digestion (Haug *et al.*, 1978). Table 3.1 shows some of the pertinent literature studies concerning

Table 3.1 Influence of pre-digestion thermal hydrolysis on sludge dewaterability.

Material	Scale	HRT	TH Temp.	Before (%)	After (%)	Change (%)	References
Mixed municipal sludge	Laboratory	15	130		27.0		Higgins *et al.* (2015)
Mixed municipal sludge	Laboratory	15	140		28.0		
Mixed municipal sludge	Laboratory	15	150		30.0		
Mixed municipal sludge	Laboratory	15	160		31.0		
Mixed municipal sludge	Laboratory	15	170		33.0		
Mixed municipal sludge	Full	22	165		31.3		Lancaster (2015)
Mixed municipal sludge	Full		165	19.0	30.0	11.0	Fjordside (2005)
Unspecified	Pilot		120	30.0	43.0	13.0	Kepp and Solheim (2000)
Mixed municipal sludge	Full	20	165		26.0		Riches *et al.* (2008)
Unspecified	Pilot		Not given	25.2	32.7	7.5	Phothilangka *et al.* (2008)
WAS	Laboratory	20	165	26	33.9	7.9	Oosterhuis *et al.* (2014)
Unknown	Laboratory			30	43.0	13	Neyens *et al.* (2004)
WAS	Laboratory	20	170	20	39.6	20	Liu *et al.* (2012a, b)
Mixed	Full-			20	30.0	10	Mills *et al.* (2013)
Mixed	Lab – WAS only			20	28.0	8	
Mixed	Pilot – ITHP			20	32.0	12	

WAS	Full	13	160–170		28–30		Nawawi *et al.* (2007)
Mixed with imports		12	165		> 30		Walley (2007)
WAS	Pilot				25		Geraats (2014)
Mixed (50:50)	Pilot				28–32		
WAS	Full			22	29–30	8	Pereboom *et al.* (2014)
Mixed	Pilot	80 h	130/30 min		31.5		Qiao *et al.* (2013)
Mixed	Pilot	80 h	130/60 min		32.3		
Mixed	Pilot	80 h	130/90 min		34.0		
Mixed	Pilot	80 h	150/30 min		46.6		
Mixed	Pilot	80 h	150/60 min		38.2		
Mixed	Pilot	80 h	150/90 min		42.2		
Mixed	Pilot	80 h	175/30 min		44.0		
Mixed	Pilot	80 h	175/60 min		47.8		
Mixed	Pilot	80 h	175/90 min		45.1		
Mixed	Pilot	80 h	190/30 min		47.8		
Mixed	Pilot	80 h	190/90 min		48.7		
Mixed	Pilot	80 h	150/90 min		46.4		

the influence of thermal hydrolysis on dewatering. Typically, thermal hydrolysis improves dewaterability of mesophilically digested sludge by up to 10 percentage points depending on influent sludge composition and dewaterer, although dewatering is significantly improved compared with no thermal hydrolysis irrespective of dewatering device.

There are many excellent reviews available on the topic of dewatering, relating to: development of rheological tools to predict performance (Stickland, 2015); quantification of (Skinner *et al.*, 2015); limitations of (Vesilind & Hsu, 1997); performance overview (Christensen *et al.*, 2015); even the influence of thermal hydrolysis itself (Neyens & Beyens, 2003) and the interested reader is referred to those texts.

According to Neyens" team (2004), advanced digestion pre-treatment technology improves dewaterability in two ways: (1) the degradation of proteins and polysaccharides from within extracellular polymeric substances (EPSs) thereby reducing its water retaining capacity (Barjenbruch & Kopplow, 2003; Chu *et al.*, 2002; Neyens *et al.*, 2004); and (2) promotion of flocculation which reduces the amount of fine flocs (Bougrier *et al.*, 2008).

The EPS matrix is very heterogeneous. A variety of polymeric materials have been found to occur in EPS, including neutral and acidic polysaccharides, lipopolysaccharides, proteins, nucleic acids and humic acids (Bura *et al.*, 1998). There is increasing evidence that the composition and properties (e.g., hydrophobicity and surface charge) of EPS is more important with respect to settleability than the amount of EPS produced. EPS constitutes approximately 80% of the composition of activated sludge (Neyens *et al.*, 2004). Hasan (2017) noted that biological sludge had approximately six times more EPS than an equivalent quantity of primary sludge. EPS has been discovered to influence flocculation, settling properties and dewaterability of sludge (Tian & Zheng, 2006). Furthermore, the composition is important, and the COD:N:P ratio of EPS influences hydrophobicity and surface charge (Bura *et al.*, 1998) and both bound water and hydrophobicity increase with depleting phosphorous concentrations. Subramanian (2004) showed that bound water content was very influential in dewaterability potential below 20% DSs; however, had little influence above that figure.

The destruction of EPS plays a key role in improving dewatering potential (Tian & Zheng, 2006). The authors, along with Neyens and Beyens (2003), proposed that increasing low levels of EPS are initially thought to aid sludge dewaterability by improving the level of sludge flocculation reducing the number of small particles present in the sludge, a factor that has been shown previously to make sludge easier to dewater. However, once a certain level of sludge flocculation has been attained, further increases in EPS become detrimental to sludge dewaterability (Neyens & Beyens, 2003). This is also proposed by Scales (personal communication). Analysing the graph presented by Houghton and co-workers (2001) this concentration is between 30 and 40 mg EPS/g suspended solids although there is no information on the quantity of the solids which is volatile, or even the composition of EPS measured.

Molokwu and Rus (2017) studied the influence of EPS on a variety of parameters known to influence both dewaterability and polymer consumption. They found a positive correlation between increasing EPS in digestate with increasing soluble phosphorous, mono to divalent cation ratio (M/D) and polymer consumption. Hasan (2017) also found a correlation with increasing protein-EPS and decreasing dewatering. Hasan also commented that thermal hydrolysis (prior to digestion) destroyed approximately 50% of protein-EPS, but after digestion EPS was similar regardless of pre-treatment. Hasan measured both tightly and loosely bound EPS in a variety of raw and digested sludges with and without thermal hydrolysis. Most of the raw sludges showed an equal split in both types; however, digestion destroyed a proportion of the loosely bound EPS. In thermally hydrolyzed digested sludge, approximately two-thirds of the remaining EPS were tightly bound.

In a review by Tian and Zheng (2006), EPSs are typically 20–55% carbohydrate, 20–80% protein with the remaining 30% comprising a mixture of humic, uronic and nucleic acids along with lipids among other minor compounds, and this is in agreement with other work (Neyens *et al.*, 2004). Nevertheless, as well as destroying EPS and enhancing compressibility of sludge, thermal hydrolysis influences a variety of other parameters which also influence dewaterability such as: viscosity, which is inversely correlated with increasing thermal hydrolysis reaction temperature (Everett, 1972; Haug *et al.*, 1978; Higgins *et al.*, 2015); increased DSs (loading rate) resulting in higher compressive yield stress (Stickland, 2015); particle size and distribution (Hendriks & Zeeman, 2009; Neyens *et al.*, 2004) and protein solubility, which influences polymer consumption (Hung-Wei *et al.*, 2014; Murthy & Novak, 1998), amongst others.

In addition, downstream anaerobic digestion further influences dewaterability of sludge by changing: viscosity which increases (Lotito & Lotito, 2014); concentration of extracellular material and other soluble microbial products (Aquino & Stuckey, 2008; Barker & Stuckey, 1999; Chu *et al.*, 2002); volatile solid concentration (Skinner *et al.*, 2015); release of nutrients and cations (Barnard & Shimp, 2013) as examples. Subsequently, some workers suggest that the act of anaerobic digestion makes rheology of digested sludge similar regardless of pre-treatment (Dawson & Sotiriadis, 2007; Xue *et al.*, 2015). There is a growing body of evidence to suggest that the aforementioned influences of anaerobic digestion of thermally hydrolyzed sludge are fundamentally detrimental with respect to dewatering potential.

In the UK, Lumb (1951) showed that thermally hydrolyzed raw sludge could achieve DSs of 52%, which is far higher than anything observed following anaerobic digestion (Table 3.2). Years later, raw thermally hydrolyzed sludge was dewatered to between 40% and 50% DS compared with sludge which had not been exposed to those conditions which peaked between 20% and 25% DSs (Sheerwood & Philips, 1970). Installation of thermal hydrolysis downstream of digestion prior to a second digestion stage has been found to improve dewaterability compared with application upstream (Rus *et al.*, 2016). Additionally, data from recent work (Svennevik, 2019b; Yang *et al.*, 2019) are complementing the work of Lumb and other early adopters (Everett, 1972, 1974; Sheerwood & Phillips, 1970), which shows that anaerobic digestion can reduce the dewatering potential of thermally hydrolyzed raw sludge by approximately 20 percentage points.

Haug and co-workers (1978) also noted that while thermal hydrolysis improved dewaterability potential with increasing reaction temperature between 100°C and 225°C, this potential was always lower than prior to digestion. Interestingly, the authors maintained that it was the dewaterability of primary, rather than activated sludge which improved and responded best to thermal processing irrespective of digestion. This may be due to activated sludge having inherently higher viscosity (Lotito & Lotito, 2014), higher quantity of EPSs prior to digestion (Neyens & Baeyens, 2003) and fundamentally different rheology (Eshtiaghi *et al.*, 2013; Markis *et al.*, 2016). As such, the findings of Everett (1972) suggesting higher temperatures of 190°C as optimal for activated sludge dewatering are typical. The application of additional heat (to 210°C) did not further improve dewaterability but allowed a shorter reaction time to be used (Everett, 1972). Everett interpreted the improvements by a reduction is specific resistance to dewatering caused by destruction of structural integrity of the micro-organisms, which is consistent with additional destruction of EPS and also further reduced viscosity (Higgins *et al.*, 2015).

Recently, Oerke and co-workers (2017) summarized the full-scale dewatering performance of over 55 plants. The breakdown of dewatering device is shown in Figure 3.2. Typically, owners prefer equipment they are familiar with, especially if they have equipment installed on other sites. With respect to performance, all facilities were over 28% DS minimum, with piston presses at 30–45% DS and plate presses above 40% DS, although the use of additives which contribute to DSs needs to be considered when reviewing plate press dewatering data. There has been a trend where belt presses have been decommissioned and replaced by centrifuges (Oerke *et al.*, 2017).

Due to changes in rheological properties, it is possible to load dewaterers at higher loading rates than for sludge which has not been conditioned. According to Panter (2009), centrifuge-treated hydrolyzed digested sludge can operate at approximately 20% higher loading rates compared with sites devoid of pre-treatment. In a joint study, Thames Water with DC Water looked at optimizing the performance of belt-filter presses. They found that loading rates could vary between 200 and 1,200 $kg/hr\text{-}m_{belt\,width}$ with no detrimental influence on either DS output or polymer consumption. This is comparable with average loading rates between 320 and 540 $kg/hr\text{-}m_{belt\,width}$ for digested sludge (Metcalf & Eddy, 2013). However, the study did highlight deteriorated fines capture at the higher loading rates, but this was managed by optimizing polymer mixing and dilution.

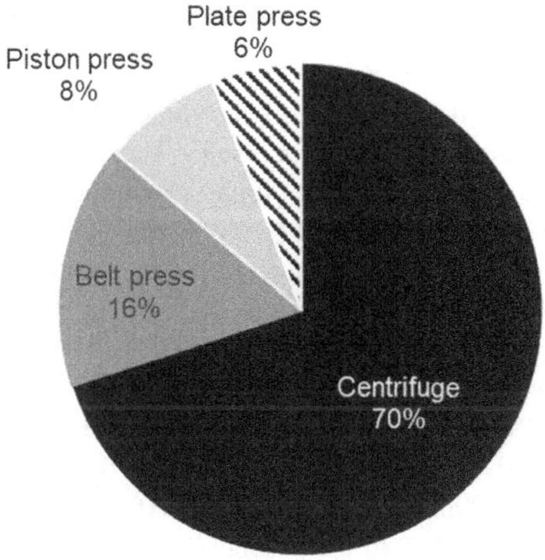

Figure 3.2 Choice of dewatering device for plants with thermal hydrolysis. Plot using data in Oerke *et al.* (2017).

However, in Oerke's summary, there was no information on the relative change in performance prior to the installation of thermal hydrolysis. Recently, a study has shown performance of full-scale plants with and without thermal hydrolysis pre-treatment based on sampling of 22 full-scale facilities some of which also had co-digestion (Svennevick *et al.*, 2019a). The authors derived an empirical relationship relating the sum of the ratio of carbon to nitrogen with ash content (termed C/N.ash) to achievable dewaterability. In this way, an accommodation could be made for both sludge types, but also biological sludge age, both influential on dewatering, but not accounted for if only volatile solids is used for prediction. Correlations were made for standard digested material, and pre- and post-digestion thermal hydrolysis. The relevant equations (taken from Svennevick *et al.*, 2019a, b) are given, namely

Cake-dried solids

$$= 5.53x + 7.14 \text{ [for non-thermally hydrolyzed sludge/biosolids]} \tag{3.3}$$

$$= 5.05x + 13.92 \text{ [for pre-digestion thermal hydrolysis]} \tag{3.4}$$

$$= 1.87(5.53x + 7.14) \text{ [for post-digestion thermal hydrolysis]} \tag{3.5}$$

where $x =$ sum of the ratio of carbon:nitrogen and percentage of material which is ash, expressed as a decimal. However, in the work, the authors found an improvement of approximately 7 percentage points when thermal hydrolysis was added prior to digestion. The empirical correlations may provide a useful tool for predicting the influence of pre- or post-treatment.

3.4 REFRACTORY COMPOUNDS FORMED DURING THERMAL HYDROLYSIS

Temperatures within a thermal hydrolysis plant produce numerous exotic compounds in the sludge being processed. These compounds are largely refractory, manifested by decreases in biogas production (Stuckey & McCarty, 1984) despite increases in soluble COD (Dwyer *et al.*, 2008; Tampio *et al.*, 2014) with increasing thermal hydrolysis temperature. In addition to temperature, the reactions producing these compounds are more likely at higher pressure and have alkaline pH optima. From approximately 140°C to 160°C, Maillard reaction end-products are created. These materials are non-enzymatically formed brown co-polymers created by complex networks of series and parallel reactions involving carbonyl and amino compounds (Maillard, 1912), such as reducing sugars and amino acids (Echavarría *et al.*, 2012). At similar and slightly increased temperatures, sugars caramelize in the absence of proteins in other non-biological reactions to create chromophore products. Therefore, sugar caramelization often results in over estimation of the Maillard reaction. Ascorbic acid oxidation is an additional mechanism resulting in the formation of colored compounds but is less relevant with respect to sewage treatment. The chromophores produced via these various mechanisms have several detrimental impacts on wastewater treatment (Batstone, 2017) as follows:

- Contribution of nutrients and COD to final effluent
- Inhibition to both anaerobic digestion and deammonification-based systems
- Interactions with polyelectrolytes potentially increasing demand
- Increased disinfection demand due to interference with UV treatment
- Fouling of microfiltration

The presence and understanding of the mechanisms of these compounds is therefore of critical interest. Most of what is currently understood about production of melanoidins formed by the Maillard reaction can be traced back to the pioneering work of Hodge (1953). First, the carbonyl group of a sugar reacts with an amino group from a protein and produces a highly unstable nitrogen substituted glycosylamine (*N*-glycosylamine) and water. Second, the carbon skeleton of the *N*-glycosylamine is then rearranged to make a more stable isomer in a process known as Amadori rearrangement (Kuhn & Weygand, 1937 cited by Hodge, 1955) and results in the formation of the corresponding fructose–amino acid complex. Next, depending on the system pH, this material is converted to either fission products (such as acetol and pyruvaldehyde), reductones or to the Schiff base of hydroxymethylfurfural or furfural. These products are then degraded to precursors of melanoidins including aldols, nitrogen-free polymers and aldehydes. Finally, these materials combined with amino compounds produce brown-colored nitrogenous polymers and co-polymers. It is only in this last step that color is produced. The pathways are shown in Figure 3.3. These materials are typically not

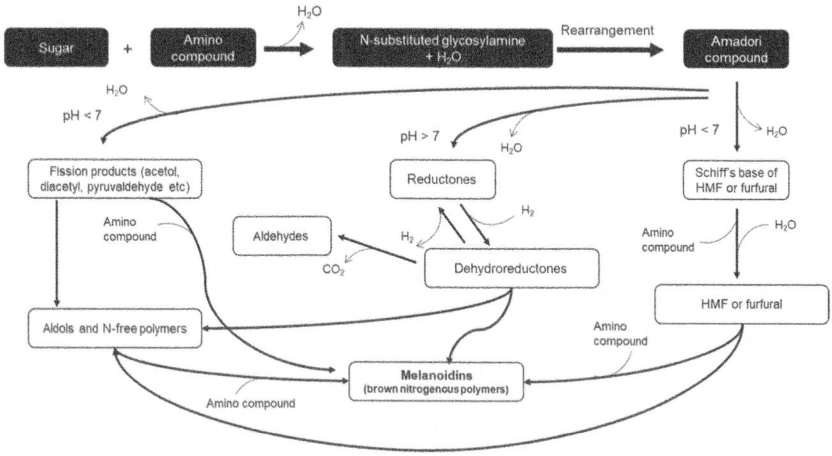

Figure 3.3 Generation of Maillard reaction products from simple sugars and proteins (adapted from Hodge, 1953; Kumar & Chandra, 2006).

biodegradable (Stuckey & McCarty, 1984) and several workers have found a reduced normalized biogas yield with thermal hydrolysis of food-waste due to Maillard and caramelization reactions (Liu *et al.*, 2012a, b; Tampio *et al.*, 2014). At slightly higher temperatures, sugars are hydrolyzed to glucose and fructose prior to decomposing to a variety of products shown in Figure 3.4.

In a very informative investigation, Ahuja (2015) analyzed the impact of various parameters on the production and type of refractory material produced during thermal hydrolysis and anaerobic digestion. In that study, material was

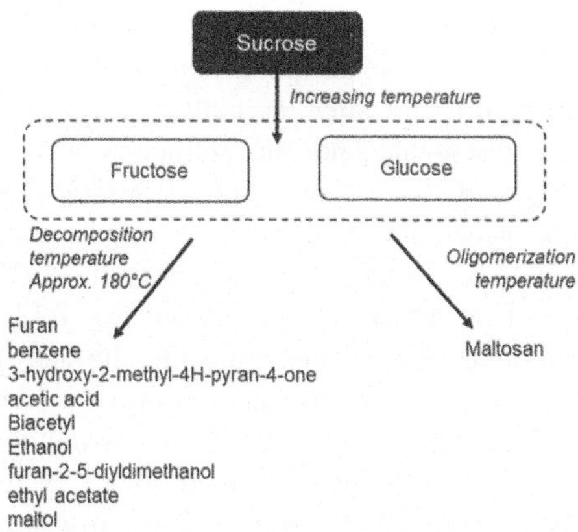

Figure 3.4 Caramelization of sugar (adapted from Ajandouz & Puigserver, 1999; Kitaoka & Suzuki, 1967; Monte & Maga, 1981).

subdivided into humic acid-like, fulvic acid-like and hydrophilic material, measured by molecular weight fraction at 45 μm, 100, 3 and 1 kDa, and at thermal hydrolysis operating temperatures between 130°C and 170°C. The interested reader is encouraged to view that work, as only a high-level summary is given here. The work showed that dissolved organic nitrogen (DON) was found to increase in a linear fashion with reaction temperature from approximately 340 to 705 mg/l. On closer inspection, this increase was due to rising levels of fulvic and hydrophilic materials, as humic-like substances showed an opposite trend. Increase in operating temperature resulted in the breakdown of humic acid materials to fulvic acids. The composition of the compounds also reduced with reducing molecular weight fraction, and this showed similar trends albeit with higher numbers between 130°C and 170°C. At 170°C, over 50% of all the DON measured was greater than 100 kDa in size. At lower temperatures, there were larger concentrations of smaller molecular weight material. Therefore, at higher thermal hydrolysis operating temperatures, the success of coagulation (see later) will be more. With respect to UVA_{254}, a noticeable trend was observed with reactor operating temperature with increases of approximately 30% and 75% at 150°C and 170°C, respectively, compared with data collected at 130°C (Ahuja, 2015). At 130°C, there was a uniform distribution across molecular weight fraction; however, at the higher temperatures this shifted toward the 0.45 μm range. Interestingly, a strong correlation was found between increasing levels of DON and UVA_{254}, to the point where they could be used as surrogate measures for each other. Ahuja (2015) also measured dissolved organic carbon (DOC) and found that it also increased with temperature. At 130°C and 150°C, levels were approximately 1,550 mg/l, but these jumped by an additional 1,000 mg/l at 170°C. The author also discovered that the DOC/DON ratio decreased with temperature, subsequently nitrogen within these materials becomes far more relevant as temperature increases.

There is a paucity of literature on how much of the processed sludge becomes refractory; however, there are some published data showing differences in effluent concentration before and after thermal hydrolysis was installed. Using this, it is possible to back-calculate the refractory content of the liquors. deBarbadillo (2016) showed that effluent refractory DON increased from 0.81 mg/l (in 2012 prior to installation of thermal hydrolysis and digestion) to 1.05 mg/l after installation. The filtrate flow accounts for 0.3% of the influent, therefore, the quantity of refractory nitrogen added was ((1.05 – 0.81)/0.3% =) 80 mg/l. Based on reported performance from the site and using stoichiometry the expected release of nitrogen from thermal hydrolysis is in the region of 2,400 mg/l. Of this, 80 mg/l has gone through the digestion and dewatering plant, and this is equivalent to a little over 3% of the nitrogen released in the digester. However, while no data on removal was reported across the digestion process, Higgins and co-workers (2017) provided data on impact of dewatering on UVA_{254}. At the typical operating temperature of the thermal hydrolysis plant in

Blue Plains, there is an approximate reduction of half of the problematic material. Unfortunately, no data were presented on dewatering setup, type or quantity of polymer used. Assuming this figure, then the refractory nitrogen produced as a fraction of total nitrogen released is between 5% and 6%. An earlier study in Australia (Dwyer *et al.*, 2008) provided similar data with which a comparable determination could be made. DON at that site increased from 1.8 to 2.8 mg/l, which based on influent loads was approximately 5% of the nitrogen released after dewatering, and making similar assumptions to the case for the facility in Washington DC, 10% of the nitrogen released in the digester. This is higher than Washington; however, the sludge in the Australian study had a far higher concentration of waste-activated sludge. Unpublished work on seven full-scale plants shows that color, DON, recalcitrant COD and UVA_{254} are approximately twice as high in plants processing mainly (80%) biological sludge compared with facilities treating mixed sludge where primary is the majority type. Two plants treating food-waste also showed higher figures for color and UVA_{254} compared with primary sludge-laden plants, even with a reaction temperature 20°C lower. However, no statistical difference was noted for non-biodegradable COD and DON.

3.4.1 Properties and types of colored refractory compounds

The chemical properties of melanoidins are akin to those of humic substances, being acidic, polymeric and highly dispersed colloids (Migo *et al.*, 1993). Melanoidins contain large quantities of nitrogen by weight. Kato and Tsuchida (1981) provided a compositional breakdown: carbon 48%; hydrogen 5.5%; oxygen 6.5% and nitrogen 40%. Wang's group (2019) report that 40–70% of nitrogen, 50–70% of potassium and 10–15% of phosphorus in sewage sludge could be transferred to these materials. A diverse group of compounds, they have been found to have high molecular weight, with 80% over 10 kDa (Liu *et al.*, 2012a, b). Although melanoidins are chemically diverse, many studies have shown them to be highly anionic (Echavarría *et al.*, 2012) with an isoelectric point as low as 2.5, which can explain how they can be removed using appropriate dewatering additives (Migo *et al.*, 1993, 1997). However, while looking at melanoidins in coffee, Echavarría's group (2012) found that anionic charge increased with molecular weight. But, increasing roasting temperatures resulted in a decrease in molecular weight. Based on that work, reducing thermal hydrolysis reaction temperature should increase molecular weight of the chromophores and increase their charge improving the ability to remove them during dewatering.

A great deal of work has highlighted the chelating properties of melanoidins, especially for iron. This property has been hypothesized to be one of the mechanisms by which melanoidins have anti-microbial properties with the ability to inhibit growth of pathogenic organisms (Morales *et al.*, 2005). However, on a treatment plant, these chelating properties have been found to render supplemental micronutrients added to a deammonification sidestream plant (treating liquors from thermal hydrolysis and digestion) ineffective (Driessen *et al.*, 2018).

Production of the recalcitrant components is directly influenced by several parameters as follows:

- Temperature
- pH
- Moisture content/water activity (not relevant to wastewater treatment)
- Types of proteins and carbohydrates and their relative molar ratio
- Biochemical agents and
- Chemical inhibitors

Some of these are routinely manipulated in the food industry to either promote or depress the production of complex colored compounds.

3.4.1.1 Temperature

Maillard himself noted that production of refractory material was directly related to temperature (1912). This finding has been replicated manifold times across several scientific fields, including that of wastewater treatment. Heat increases the rate of chemical reactions in accordance with Arrhenius kinetics and accelerates the evaporation of water. In the food industry, evaporation further concentrates the melanoidins. It is also well documented in that industry that pressure cooking of food, thus increasing boiling point, encourages the Maillard reaction. Similar to temperature, time of heat treatment also influences production of the colored compounds (Labuza & Baisier, 1992). In addition to production of melanoidin, temperature also increases the activity between proteins and sugars. Also, the types of compounds generated are also affected by temperature. In a detailed study, Ahuja (2015) showed that the molecular weight of the color-harboring compounds increased according to temperature between 130°C and 170°C. At 130°C, molecular weight distribution was approximately uniformly dispersed between >300, 300, 3 and 1 kDa. However, this shifted to a linear increase with temperature at 170°C with emphasis on higher molecular weight compounds. This is mirrored in the work of Oracz and Nebesny (2019) looking at roasting cocoa beans. On closer inspection, DON, fulvic acid and the hydrophilic fraction of DON all increased while quantity of humic acid decreased with temperature.

Therefore, reducing temperature can alleviate potential issues with refractory compound formation. Dwyer's team (2008) showed that production of color materials dropped significantly with temperature (from 12,677 to 3,837 mg/l PtCo – a reduction of 70% – by reducing temperature from 165°C to 140°C), with no subsequent reduction in biogas or dewaterability. The work of Dwyer's group has been noted in various other studies (Ahuja, 2015; Higgins *et al.*, 2017; Liu *et al.*, 2012a, b; Svennevick *et al.*, 2019b; Wilson & Novak, 2009).

3.4.1.2 pH

Both Maillard and caramelization reactions have alkaline pH optima, as found in digesters following thermal hydrolysis, which further exacerbates the issue. Lowering pH has been used with success with respect to reducing production of refractory material (Nursten, 2005).

3.4.1.3 Type of protein and sugar

Liu's research team (2012a, b) showed that UVA absorbance was highly influenced by the type of substrate processed with significantly higher UVA absorbance noticed with kitchen and food-wastes compared with activated sludge (Liu *et al.*, 2012a, b). Measurements of carbohydrates and proteins present did not fully coincide with the UVA absorbance measurements. The same study showed that while biogas production increased for activated sludge by over 30%, it decreased for kitchen and food-wastes by 8% and 12%, respectively. Therefore, production of melanoidin products and caramelans needs to be considered when reviewing the potential for food-waste co-digestion with thermal hydrolysis. Subsequent work has looked at the melanoidin production of different types of sludges (Svenevick, *in press*). That study has shown that color, DON, UVA_{254} and refractory COD all increase according to increasing activated sludge levels.

3.4.2 Removing refractory compounds

The structure and characteristics of melanoidins are still not fully understood and this has prevented development of formal removal procedures (Nguyen *et al.*, 2010; Satyawali & Balakrishnan, 2008). However, removal of refractory compounds by prevention or direct treatment has been successful. Preventative methods involve manipulation of the sludge feedstock prior to thermal hydrolysis, or adjusting thermal hydrolysis operating conditions – although there have been questions over the efficacy of this abatement method, whereas direct treatment involves using purpose-built equipment to remove the produced material. Alternatively, newer approaches involve the recovery of the materials (Ward *et al.*, 2018) as they have many beneficial characteristics.

3.4.2.1 Prevention

If sludge is preconditioned prior to thermal hydrolysis it may prevent the occurrence of melanoidins. The use of chemicals to start hydrolysis and solubilization while reducing viscosity prior to heat treatment, may allow the benefits of thermal hydrolysis to be realized at lower temperatures, which would decrease the production of refractory components. Although not done with the intent of reducing the production of refractory compounds, Abelleira's team (2012) studied the combination of hydrogen peroxide pre-treatment followed by thermal hydrolysis and subsequent anaerobic digestion. The aim was to take advantage of the synergy between peroxidation and direct steam injection. By adding peroxide, the optimum temperature of the thermal hydrolysis plant dropped to 115°C, at which temperature melanoidin production is relatively low. Combined treatment outperformed thermal hydrolysis by itself in biogas production and especially dewatering. However, the optimum conditions for peroxide addition were not considered viable at industrial scale. In spite of this, the work highlights an interesting area for future exploitation.

3.4.2.2 Coagulation

When treating alcohol distillery wastewater with high levels of melanoidins, Migo and co-workers (1993, 1997) observed the impact of inorganic ions of $FeCl_3$, $AlCl_3$ and polyferric sulfate on residual turbidity. They found an optimal addition dose of approximately 0.04 M of the compounds where residual turbidity was reduced to a point of no detection. The authors looked at pH to elucidate on which species were influential. The best results were observed at pH 2–4, under which conditions the dominant species were unhydrolyzed Fe^{3+}, $Fe(OH)_2^+$ and $Fe(OH)_3$. Based on this, the authors described the mechanism of melanoidin removal as follows: (1) charge neutralization of the negative melanoidin colloids by Fe^{3+} and $Fe(OH)^{2+}$; (2) adsorption of $Fe_2(OH)_2^{4+}$ onto the surface of the colloidal particles to enable interparticle bridging and (3) sweep flocculation (O'Melia, 1972) by the Fe $(OH)_3$ precipitates.

However, higher doses of the chemicals resulted in the return of the turbidity to similar levels prior to addition, consistent with an overabundance of positive charges which would result in deflocculation. It is not known how much of the removal is down to chemical addition, or how much due to pH adjustment, as that is a known remover of melanoidin-induced color in its own right. The authors concluded by stating which of the chemicals worked best and their optimal pH. $FeCl_3$, $Fe_2(SO_4)_3$ and polyferric hydrosulfate (PFS) removed over 95% of the color present with pH optima between 3 and 4.5 for the first two chemicals, and approaching neutral pH for PFS. $Al_2(SO_4)_3$ removed approximately 80% at a slightly acidic pH. Calcium oxide and chloride removed approximately 75% and 45% of the color under strong ($pH > 13$) alkaline conditions (Migo *et al.*, 1997). The influence of pH change due to addition of coagulants should not be underestimated, as it may consume alkalinity required by downstream deammonification processes, thereby incurring an additional operating cost in supplementation of further alkalinity at that stage (Wang *et al.*, 2018).

A team lead by Nguyen (Nguyen *et al.*, 2010) looked at a variety of ways to remove melanoidin-caused color from industrial wastewater. As part of this study, the authors looked at the following coagulants: alum, lime, ferric chloride, aluminum chlorohydrate (ACH) and polydiallyldimethylammonium chloride (more commonly known as polyDADMAC). They were all tested individually and in various combinations. The industrial wastewater contained 30,000–37,000 color units (Pt–Co), between 8,000 and 10,000 DOC (mg/l) UVA_{254} 0.11–0.16 for a 1–1,000 dilution, and SUVA readings of 1.38–1.40 l/m/mg. They found that the best removal rates were due to the addition of lime at a dose of 20 g CaO/l and this corresponded to a removal of 60% of the color. However, this significantly increased pH (which would remove color by itself) and produced a sludge which would most likely be identified as a special waste due to its high pH (>12). By contrast, polyDADMAC showed only a modest removal of 30% when added at the optimal dose of 1.2 g/l. The addition of alum to polyDADMAC reduced this further to approximately 20% while the addition of ACH increased performance to 40%. ACH, alum (both dosed at 200 mg Al^{3+}/l) and ferric chloride shared limited success with removal rates between 15% and 20%. The addition of ferric chloride reduced pH to below 3 and resulted in a voluminous production of sludge (Nguyen *et al.*, 2010). The authors concluded that, in spite inferior performance, a combination of ACH with polyDADMAC (at a dose of 0.96 g/l ACH with 1.2 g/l polyDADMAC) was the most suitable option with respect to coagulation as it did not significantly alter pH, required a much lower dose than lime, and did not result in the production of large quantities of sludge.

Wilson and co-workers (2011) found that they could manipulate the measured organic nitrogen in centrate by treating with ferric chloride and an unspecified cationic polymer. With neither present, centrate contained 260 mg N/l. This reduced to 140 and 100 mg/l for addition of cationic polymer at optimal dose (not specified), and 50% cationic polymer with 0.05 g ferric chloride/g TS, respectively. When the ferric dose was doubled in conjunction with 50% cationic polymer, nitrogen was not detected. In a later work, coagulation was removed up to 16% of soluble COD and approximately 95% DON and between 17% and 29% UVA_{254}, although no data were presented on types or concentrations of coagulants (Pace *et al.*, 2018). Zhang *et al.* (2018) used coagulants to remove large colloidal material which reduced inhibition of downstream deammonification. Specifically, the addition of 2 kg/TDS polyDADMAC with 10 kg/TDS FLOPAM enabled activity potential of aerobic ammonia-oxidizing bacteria to be at 100%.

3.4.2.3 Ozone

Outside of the water industry, ozone has been routinely used for the destruction of melanoidins (Satyawali & Balakrishnan, 2008). Nguyen *et al.* (2010) tested an ozone concentration of 7 mg/l alcohol distillery wastewater with a reaction time of 45 min. This work was done alongside a study on coagulants and ultrafiltration. Compared with coagulation, ozone treatment was found to be far more efficient at removing color, DOC and COD with removal rates of 86%, 71% and 51% compared with 40%, 12% and 8%, respectively. Both coagulation and ozone shared similar removal of UVA_{254} with only 30% removed. When coagulation was combined with ozonation elimination of UVA_{254} increased to 43% although removal of the other compounds remained unchanged (Nguyen *et al.*, 2010). Three-dimensional excitation and emission matrices showed that the coagulants removed only a small quantity of fulvic and humic acid type materials, whereas ozonation almost totally degraded these compounds and other soluble microbial products. In a later work, Pace's team (2018) found that ozone was effective at reducing sCOD (5–7%), DON (60–95%) and UVA_{254} (3–6%) in a dewatering sidestream from thermally hydrolyzed digested sludge. Unfortunately, there was no information available on dose or experimental setup other than doses between 1 and 8 mg/l ozone were tested.

3.4.2.4 Chemical inhibitors

Sulfur dioxide is unique in its ability to inhibit the Maillard reaction and can be applied as a gas or in solution as sulfite or bisulfite. Sulfur dioxide is not only capable of partially bleaching chromophores that have already formed, but also inhibits color formation at the beginning of the reaction. Sulfur dioxide binds with glucose to form hydroxy-sulfonate and similar analogues from which it can be reversibly released. This blocks the carbonyl group of the sugar, making it unavailable for interaction in Maillard reactions (Eskin *et al.*, 2013). Aspartic and glutamic acids have also been used to prevent Maillard reactions.

3.4.2.5 Filtration

Nguyen *et al.* (2010) used ultrafiltration using 100 and 30 kDa molecular weight cut-offs with a trans-membrane pressure of 220 kPa. They showed 52% and 66% color removal at 100 and 30 kDa, respectively, and performed better than a number of coagulants the team also studied. However, severe membrane fouling made the permeate flux extremely low. Flux was improved by addition of ACH/polyDADMAC, but this did not enhance color removal and membranes suffered from fouling due to high levels of organics in the wastewater.

3.4.2.6 Bacterial decomposition

There are various fungi, algae and facultative bacteria capable of melanoidin decolorizing activity (MDA). A variety of facultative anaerobes exhibiting MDA have been identified which can remove approximately a quarter of melanoid present (Ohmomo *et al.*, 1988). This would explain the work of Dwyer's group (2008), and later Higgins" team (2017) who demonstrated a reduction in various parameters associated with refractory components following digestion. Members of the *Bacillus* species have been found to be successful in the removal of artificial melanoidins (Kumar & Chandra, 2006). *Bacillus thuringiensis* successfully removed all the color present followed by *Bacillus brevis* and

Bacillus sp. It was easier to remove materials which contained glucose as the primary source, rather than compounds containing both carbon and nitrogen. It was found that color removal could be improved by fortuitous metabolism by the addition of 1% glucose as a supplementary carbon source (Kumar & Chandra, 2006). The microbial decolorization of melanoidins is twofold, with certain organisms targeting smaller molecular weight melanoidins (such as W-NS strain) and others specializing against larger molecular weight melanoidins, such as *Basidiomycetes* and *Ascomycetes* (Ohmomo *et al.*, 1988). In all instances, supplementation with either nitrogen or carbon source is almost always necessary because the microbial species are not able to utilize the refractory material as the sole carbon source. It is suggested that more research be conducted in this area to optimize the role of digestion itself as part of a decolorization abatement strategy.

3.4.2.7 Advanced oxidation process

Nguyen *et al.* (2010) looked at wastewater which was previously coagulated using ACH and polyDADMAC and settled for 2 hr. Filtered supernatant (5 μm) then treated by UVC/H_2O_2 with a fluence of lamp at 12.95 mJ/s and a peroxide dose of 5.3 g/l. The advanced oxidation process increased overall color removal from 40% (due to the coagulants) to nearly 60% at 1 hr. A further hour provided little additional benefit. The authors suggested that this method was not economical at the doses used in the study. The results were not as successful as previous authors (Dwyer *et al.*, 2008; Dwyer & Lant, 2008); however, in those studies the color levels were significantly lower to begin with.

3.4.2.8 GAC

Various workers have looked at the efficacy of granular-activated carbon (GAC) to adsorb melanoidins and similar molecules. Simaratanamongkol and Thiravetyan (2010) studied the decolorization of melanoidins from a synthetic sugar syrup solution and found typical behavior which could be modelled by standard Langmuir and Freundlich isotherms in endothermic reactions (Kaushik *et al.*, 2017; Simaratanamongkol & Thiravetyan, 2010). The workers found increasing removal potential between temperatures of 5°C, 25°C and 60°C, of 200, 208 and 232 $mg_{removed}/g$ adsorbate. The authors concluded that 8.33 kg of GAC (derived from bagasse bottom ash) per m^3 wastewater could reduce melanoidin concentration from 100 to <10 mg/l.

3.4.2.9 Recovery

Melanoidins contain a wealth of attractive properties which makes them candidates for recovery. Previously, capillary zone electrophoresis has been studied as a means to recover these materials (Morales, 2002). A less complicated method of recovery is adsorption/desorption using commercial GAC. Kaushik *et al.* (2017) successfully recovered melanoidins from sugar-distillery wastes using activated unburnt activated carbon and commercial activated carbon with recovery rates of 78% and 80% respectively by using 25% pyridine solution to enable desorption. In a later work, Ward *et al.* (2018) have investigated the potential for electrodialysis to concentrate nutrients in wastewater to maximize recovery potential. Using a 30 cell-pair reactor with 7.2 m^2 active surface membrane area, the study managed to concentrate ammonia and potassium to approximate concentrations of 7,100 and 2,500 mg/l, respectively. Power consumption was approximately 5 kWhr/kg N recovered which was considered comparable with other standard methods of nutrient treatment.

3.4.2.10 Discussion

Refractory compounds produced mainly as Maillard reaction end products and to a lesser extent from caramelization reactions must be accounted for when considering the installation of thermal hydrolysis. It appears that sludge type is very influential, with the production of recalcitrant materials being closely correlated with the quantity of activated sludge present. Thermal hydrolysis temperature shows a strong link with production of these compounds and it is possible to reduce potential generation significantly by operating at lower reaction temperatures, although this may come at the expense of reduced biogas production and dewaterability. Increasing reaction temperature also produces more nitrogen-containing compounds, and a trend is observed whereby humic acid-like materials are converted to fulvic acid-type compounds as temperature surges. A fraction of recalcitrant materials (<25%) will be degraded during the anaerobic digestion process by facultative bacteria with MDA properties. During dewatering, it is possible to remove a further fraction by coagulation with materials with cationic charge. Lime produces the best results but changes the pH of the biosolids fundamentally and generates an abundance of additional mass. A better alternative is to use a combination of ACH and polyDADMAC which can be done at lower doses and remove up to half of the material left. Post dewatering, ozonation has proved to be highly effective means of extraction, although membranes and GAC have also shown positive results. However, although problematic in the wastewater industry, melanoidins carry numerous positive characteristics including: antimicrobial, antioxidant, anticancer and detoxifying activity. Subsequently, they have a high value, and rather than destroying them, it may be more beneficial in the longer term to look at ways of recovering them as a valuable resource. Although the exact composition of melanoidins is unknown, their production, manipulation and inhibition are fundamentally important in the food industry which has been researching into these compounds for over 100 years. The interested reader will learn a great deal from exploring the exhaustive work from that industry.

3.5 IMPACT ON AMMONIA TOXICITY DURING ANAEROBIC DIGESTION

It is well understood that free unionized ammonia NH_3 (also known as free ammonia – FA) controls the upper pH limits of anaerobic digestion. However, there is evidence to suggest that charged ammonium (NH_4^+) may inhibit methane synthesizing enzymes directly (Chen *et al.*, 2014). Therefore, the loading rates to digesters which are fed thermally hydrolyzed sludge are limited by the expected total ammonia nitrogen (TAN) and FA concentrations in the digester. Therefore, digesters are typically fed at 10% DSs, which may be slightly higher for sludges rich in primary sludge, and lower for those where there is an abundance of activated sludge.

As thermal hydrolysis allows an increase in loading rate due to altered rheology (Dawson & Sotiriadis, 2007), increases solubility of proteins (Haug, 1977; Hung-Wei *et al.*, 2014; Li & Noike, 1992), and improves the breakdown of those proteins (Bougrier *et al.*, 2008), an increase in ammonia and also alkalinity is noted, resulting in a pH rise. Subsequently, the operating pH of a mesophilic digestion plant preceded by thermal hydrolysis is routinely between 7.5 and 8. However, increasing pH (and also temperature) shifts the equilibrium position away from ammonium to its free state. Having no charge, FA diffuses easily into a cell and once there, ionizes to form ammonium resulting in an intracellular pH imbalance which in turn, stimulates a K^+ /H^+ anti-porter mechanism. This has been evidenced by the work of Sprott *et al.* (1984) who found a correlation between increasing ammonia concentration and depletion of potassium. Interestingly, this may enhance dewaterability characteristics of the digested sludge by altering the mono- to divalent cation ratio according to the divalent cation bridging theory for biofloc formation (Higgins & Novak, 1997; Murthy & Novak, 1998).

Such is the perceived influence of ammonia toxicity on thermal hydrolysis, that it is currently the rate-limiting design consideration. Subsequently, it is necessary to dilute hydrolyzed sludge prior to digestion to reduce its influence (as shown earlier in Figure 3.3). Typically, this limits the DS content of the feed to approximately 10% DS for mixed sludges at full-scale, however this can drop for systems only digesting waste activated sludge due to its higher nitrogen content.

Early work has suggested an FA concentration of 150 mg/l to be inhibitory to anaerobic digestion (McCarty & McKinny, 1961). Figure 3.5 shows the influence of sludge type and thermal hydrolysis on the expected FA concentration, calculated by stoichiometry using elemental composition data for primary and activated sludge presented before (Barber *et al.*, 2015), and volatile destruction data typical of full-scale application.

Figure 3.5 suggests an FA concentration for standard mesophilic digestion of between 20 and 25 mg/l, and a concentration of between 150 and over 200 mg/l depending on the sludge type and how much is exposed to thermal hydrolysis.

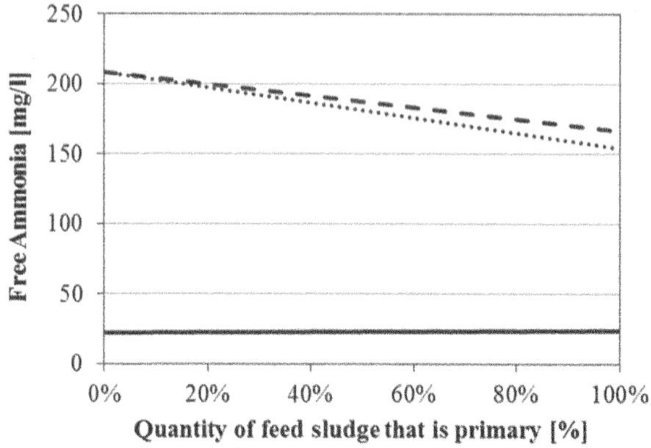

Figure 3.5 Influence of sludge type and thermal hydrolysis on expected FA concentration in anaerobic digestion. *Key*: Solid line (no thermal hydrolysis, pH 7.3, digestion temperature 36°C); dashed line (thermal hydrolysis of both primary and waste activated sludge, pH 7.8, digestion temperature 40°C); dotted line (thermal hydrolysis of only waste activated sludge, same conditions as for thermal hydrolysis of all sludge).

These figures correspond to between 2,400 and 3,500 mg/l TAN. These findings are similar to those summarized by Rajagopal *el al.* (2013) who reported on the anaerobic digestion of chicken manure and found no toxicity to methanogenesis up to 250 mg/l FA in digesters fed at 10% DSs. Oosterhuis *et al.* (2014) also found no influence of TAN of 4,000 mg/l with lab-scale thermal hydrolysis experiments. If pH is slightly higher, as witnessed on some thermal hydrolysis facilities, the FA could increase to 300 mg/l for a feed consisting of exclusively waste-activated sludge. As these figures are routinely observed at full-scale on well-performing facilities, they do not coincide with the conclusions of McCarty and McKinny from 1961. In fact, looking through the literature related to ammonia inhibition reveals a wide range of concentrations reported as being inhibitory. Other than pH, ammonia inhibition appears to be influenced by acclimation (Zhang *et al.*, 2014); bacterial population (Banks *et al.*, 2012; Fotidis *et al.*, 2013; Wiegant & Zeeman, 1986; Zhang *et al.*, 2014); temperature (Angelidaki *et al.*, 1993); nutrient content (Banks *et al.*, 2012) and seed inoculum (Yenigün & Demirel, 2013) amongst others, which makes comparison with specific data difficult. Subsequently, in a review article investigating the impacts of toxicants on anaerobic digestion, Chen and co-workers (2014) described a study which concluded that a threshold for FA which could be tolerated to be as high as 620 mg/l, far in excess of the typical figures experienced in digestion following thermal hydrolysis.

With respect to thermal hydrolysis, there is debate on how ammonia influences performance. A wide body of literature suggests that acetoclastic archaea are more sensitive to FA inhibition than either acetate-oxidizing or hydrogenotrophic methanogens (Angelidaki *et al.*, 1993; De Vrieze *et al.*, 2015; Fotidis *et al.*, 2013; Koster & Koomen, 1988; Wang *et al.*, 2015; Wilson *et al.*, 2008; Zhang *et al.*, 2014). Consequently, it is expected that digesters fed thermally hydrolyzed sludge will provide conditions to selectively favor those trophic groups (De Vrieze *et al.*, 2015; Ho *et al.*, 2013), although as will be shown later in this chapter, this is not necessarily the case. Various attempts have been made to account for a shift in bacterial population from acetophiles to hydrogen consumers and acetoclasts in response to increasing ammonia concentrations in predictive modelling (Haile *et al.*, 2015; Ho *et al.*, 2013; Wett *et al.*, 2014) although none have been used to predict future performance. However, in contrast, a thorough study by Wiegant and Zeeman (1986) showed that toxicity inhibited hydrogen utilizers more than their acetogenic counterparts up to at least 4,500 mg/l TAN.

One study looking at ammonia inhibition of methanogenesis only found severe inhibition at 10,000 mg/l TAN, with a negligible influence on methanogen mrcA (the gene coding for the alpha subunit of methyl-coenzyme M reductase) between 3,000 and 7,000 mg/l TAN (Zhang *et al.*, 2014). The study found stable communities of Methanosaetaceae, which does not use hydrogen as an electron donor, up to 7,000 mg/l TAN; however, above this level Methanobacteriales were more resistant. By contrast, Schnürer's group (2008) noted a shift away from acetoclastic methanogenesis to syntrophic acetate oxidation beginning at 3,000 mg/l TAN by measuring radioactive isotopes. However, this shift coincided with a reduction in methane and biogas yield.

In an other work, Wilson's coterie (2012) investigated the influence of FA on K_s for unionized acetic acid to explain why volatile fatty acids were observed in digester effluents. Although they looked at conditions with and without thermal hydrolysis, a closer analysis of the data shows that increasing ammonia concentration increased K_s irrespective of the presence or absence of thermal hydrolysis. The influence of FA on K_s derived from data presented by Wilson *et al.*, is shown, namely

$$K_{s[HAc]} = -2.077 \ln(\mathrm{FA}) + 12.637 \tag{3.6}$$

where $K_{s[HAc]}$ = half saturation constant for unionized acetic acid (mg/l) and FA = free ammonia concentration (mg/l). This compares with previous work showing K_s of 0.128 mg/l (Kus & Wiesmann, 1995) for a system without thermal hydrolysis – equivalent to an FA of about 410 mg/l HAc.

A study by Ngwenya's group (2015) showed that reducing digestion retention time could alleviate ammonia toxicity, implying that, following thermal hydrolysis, degradation of carbohydrates was occurring in preference to the breakdown of proteinaceous materials as would be expected from the Gibbs free energy for the relevant reactions (McCarty, 1971). TAN figures increased with retention time from 1,500, 2,500 and 3,000 mg/l at retention times of 10, 15 and 18 d with adjusted loading rates of 5.5 kg VS/m^3 d each. Alkalinity followed a similar trend and pH increased from 7.4 to 7.7 and 7.8 from the shortest to longest retention time. Although not presented, the corresponding calculated FA concentrations are 56, 180 and 266 mg/l. No statistical difference was observed on biogas production between 10, 15 and 18 d suggesting no adverse influence of FA on performance between 56 and 266 mg/l. The work implies an optimal hydraulic retention time by when most carbohydrate and lipid material have been degraded but little protein. Extending retention time would improve protein degradation resulting in higher pH and FA and eventually lower biogas yield.

One way to relieve ammonia toxicity is to feed carbon-rich co-substrates to lower pH (Chen *et al.*, 2014; Kayhanian, 1999). With thermal hydrolysis, fats oil and grease supplemented as 50% COD caused a drop in pH of 0.1 units (Ngwenya *et al.*, 2015). A similar drop was noted with the addition of food-waste, albeit at a rate of only 25% additional COD load (Barber *et al.*, 2015). From equilibrium calculations, a reduction of 0.1 pH will reduce FA by approximately 20% at the typical TAN concentrations and digester operating conditions under thermal hydrolysis. The relatively low drop in pH units following thermal hydrolysis may be due to improved biodegradability of the intermediate products of anaerobic digestion demonstrated by data showing fatty acid intermediates contribution of only 2–4% of the effluent soluble COD following co-digestion with thermal hydrolysis (Barber *et al.*, 2015).

An easier approach to reduce ammonia toxicity is simply to reduce digestion temperature. Most digesters following thermal hydrolysis are operated at approximately 40°C which is the optimal temperature for digestion when hydrolysis is rate limiting (Tong *et al.*, 1991). Assuming thermal hydrolysis circumvents this and digester operating temperature is dropped to that typical of mesophilic digestion then FA would drop by 25–30% under other operating conditions being the same. Wilson *et al.* (2008) demonstrated negligible difference in digester performance between 35°C and 42°C, supporting evidence that FA concentration is not overly influential. Interestingly, the optimum methane production of 0.4 l CH_4/g VS fed with VSR in excess of 57% was noted at 53°C in that study. Based on typical concentrations, this is equivalent to a FA concentration of >250 mg/l. The authors stated that no signs of inhibition were observed even when ammonia was supplemented such that concentration was 2,900 mg/l TAN at pH 7.8. That work was in agreement with earlier studies on high DS digestion (Lay *et al.*, 1998) which looked at the influence of both pH (between 6.5 and 9) and ammonium-nitrogen (between 100 and 6,000 mg/l). Using the Gompertz relationship with gathered data, Lay's team proposed that it was ionized ammonium not FA which influenced biogas at concentrations between 1,670 and 3,720 mg/l (highly relevant to thermal hydrolysis systems) and that resulted in a 10% decrease in biogas production. FA only became an issue at levels in excess of 500 mg/l (in accordance with findings of Wilson *et al.*, 2008), after which a dramatic shock was noted, and this was equivalent to approximately 6,000 mg/l TAN. Observation of the data presented on full-scale plants showing high levels of performance (Lancaster, 2015; Merry & Oliver, 2015), processing DSs well above 10% (Chauzy *et al.*, 2008), and at the laboratory scale where effluent is almost free of biogas precursors (Barber *et al.*, 2015), along with the findings of both Wilson's (2008) and Lay's (1998) groups and others (Liu & Sung, 2002; Oosterhuis *et al.*, 2014) suggest that inhibition due to FA at the typical concentrations observed in digestion following thermal hydrolysis is far less influential than other parameters such as sludge composition and rheological properties. Subsequently, this also suggests that pre-digestion dilution to minimize ammonia toxicity is being conservatively managed and that higher loading rates are possible assuming ammonia is the rate-limiting design parameter.

3.6 EMERGING CONTAMINANTS

There is growing evidence of, and scientific interest in the accumulation of a variety of xenobiotic compounds in the environment. There are exhaustive lists of chemicals and some of the main ones are: pharmaceuticals, personal care products, illicit drugs, hormones and steroids, polychlorinated naphthalenes, perfluorochemicals, polychlorinated alkanes, synthetic musks, quaternary ammonium compounds, pesticides, engineered nano-materials, veterinary products and other industrial compounds and by-products. Their fate across wastewater treatment and anaerobic digestion continues to be studied by various authors, and more recently the influence of thermal hydrolysis has also been studied.

Table 3.2 Influence of MAD and with thermal hydrolysis (TH-MAD) on removal of a variety of emerging contaminants. Adapted from Taboada-Santos *et al.* (2019).

Compound	MAD (% Removal)	TH-MAD (% Removal)
Naproxen	100	100
Ibuprofen	**70**	**80**
Diazepam	70	70
Roxithromycin	65–80	65–80
Citalopram	60	60
Galaxolide	60–70	60–70
Tonalide	**60**	**40**
Trimethoprim	40–50	40–50
Fluoxetine	25–40	25–40
Carbamazepine	10–20	10–20
Oestrone	10–20	10–20
17β-oestradiol	10–20	10–20

Pei *et al.* (2016) compared ozone and thermal hydrolysis pre-treatment processes combined with mesophilic digestion against a control with no pre-treatment for the digestion of pharmaceutical sludge. The authors found a poor degradability for the pharmaceutical waste and municipal sludge for the control with 33% and 42% volatile solid destruction respectively. Ozonation improved biodegradability for both to the mid-50s range while thermal hydrolysis improved performance to above 60% volatile solid destruction.

In a later work, Taboada-Santos *et al.* (2019) studied the fate of a wide variety of xenobiotics which had been processed by thermal hydrolysis (at 170°C for 20 min) and subsequent mesophilic anaerobic digestion (MAD). Eighteen compounds were studied including: three musk fragrances, galaxolide (HHCB), tonalide (AHTN) and celestolide (ADBI); three anti-inflammatories, ibuprofen (IBP), naproxen (NPX) and diclofenac (DCF); four antibiotics, sulfamethoxazole (SMX), trimethoprim (TMP), erythromycin (ERY) and roxithromycin (ROX); four neurodrugs, fluoxetine (FLX), carbamazepine (CBZ), diazepam (DZP) and citalopram (CTL); one endocrine disrupting compound, triclosan (TCS) and three hormones, estrone (E1), 17β-estradiol (E2) and 17α-ethinylestradiol (EE2) (Taboada-Santos *et al.*, 2019). The authors looked at destruction both across thermal hydrolysis itself and after digestion. They found that thermal hydrolysis resulted in a breakdown of between 5% and 20% of the compounds due to thermal instability. However, when the additional impacts of digestion were accounted for, no statistical difference was noticed in the breakdown of the compounds between the control and test configurations for almost all the compounds tested (Taboada-Santos *et al.*, 2019). Only NPX was totally removed, and it was removed from both test and control reactors. The only statistically different results observed were for the removal of IBP which was approximately $<10\%$ better with thermal hydrolysis pre-treatment, and AHTN removal of which was reduced by over a third with thermal hydrolysis.

3.6.1 Perfluorinated chemicals

Perfluorooctane sulfonic acid and perfluorooctanoic acid (PFOA) have been attracting much interest in recent years as harmful xenobiotic contaminants. These materials are widely used in surface treatments of carpets, textiles, leather, paper and cardboard, to give them fire-retardant characteristics, and used as a surfactant in fire-extinguishing foams. They have been found to bio-accumulate and cause toxicity in the environment. Their properties as fire retardants make them very thermally stable, subsequently temperatures in excess of 1,100°C are required to destroy them. Subsequently, thermal hydrolysis with or without anaerobic digestion will have little influence other than concentration of these materials into digested sludge due to enhanced breakdown of organics.

Currently, only advanced oxidation technologies, including photocatalysis, UV photolysis, sonochemical decomposition and electrochemical decomposition, are currently the only available methods for destruction of these components (Liu *et al.*, 2012a, b). However, one international study found that using persulfate $S_2O_8^{2-}$ heated to 85°C was effective at destroying PFOA described by pseudo-first-order kinetics. The degradation was more favorable at higher pH. Potentially, the solubilization and heating afforded by thermal hydrolysis in conjunction with the use of oxidation agents such as persulfate may be an avenue for the breakdown of perfluorinated and similar compounds.

3.7 IMPACT ON MICROBIAL COMMUNITY

A great deal of work has been undertaken to investigate the microbiome with both meso- and thermophilic digestion (Demirel & Scherer, 2008). More recently, this work has included the impact of thermal hydrolysis both at laboratory and full-scale. In principle, pre-treatment is expected to change the microbial community, due to sterilization of the feed sludge, but also due to a change in chemistry, especially higher levels of FA experienced in the subsequent digester (Sprott & Patel, 1986). The toxicity of ammonia is described earlier; however, it is well documented that hydrogenotrophic methanogens are less sensitive to ammonia inhibition than their acetoclastic counterparts (Demirel & Scherer, 2008). However, *Methanosarcina* sp., which can use both acetate and hydrogen as electron donors, differs from other acetoclastic methanogens due to its morphology which affords them a level of protection (Gagliano, 2014). Methanogens such as *Methanosaeta* sp. – a strict acetophile – are prone to a toxic response due to their large surface area, and therefore diminish with increasing FA. Subsequently, a shift in methanogens from acetate to hydrogen utilizers has been observed in digesters following thermal hydrolysis (Chen *et al.*, 2018, 2019; Ennouri *et al.*, 2016; Tong *et al.*, 2019).

A comprehensive Danish study studied the microbiology of 32 full-scale anaerobic digesters over 6 years using 16S rRNA gene amplicon sequencing (Kirkegaard *et al.*, 2017). The sample set included both meso- and thermophilic digestion plants with retention times between 10 and 55 d and two with Cambi thermal hydrolysis pre-treatment at Fredericia and Næstved. The acetoclastic methanogen *Methanosaeta* dominated the sequencing libraries of the mesophilic reactors with between 60% and 80% of the reads, and this was followed by a variety hydrogenotrophs such as *Methanolinea*, *Methanospirillum*, *Methanobrevibacter* as well as *Candidatus Methanofastidiosa* (Kirkegaard *et al.*, 2017). However, in contrast to expectations (Gagliano, 2014), the digesters with thermal hydrolysis pre-treatment were also dominated by *Methanosaeta* (83–87%) which does not use hydrogen as an electron donor, suggesting that FA was not limiting performance at neither Fredericia nor Næstved. Using equation 4.4, typical FA concentrations would be approximately 50, 165 and 200 mg/l for mesophilic, thermophilic and mesophilic digestion, respectively, with thermal hydrolysis. Therefore, it is a surprise that thermal hydrolysis pre-treatment resulted in a similar archaeal composition to mesophilic digestion rather than thermophilic. However, in a separate study, ammonification of proteins did not negatively affect the methanogenic activity during digestion with methanogens being able to thrive under TAN of 4.2 g/l prior to significant inhibition (Chen *et al.* 2008). In Kirkegaard's group's study (2017) a major difference in archael population was related to the well-known hydrogen consuming *Methanobrevibacter* which was present under mesophilic conditions but was conspicuously absent with thermal hydrolysis.

Following *Methanosaeta* (Kirkegaard *et al.*, 2017), the next most prominent methanogen found in digestion with thermal hydrolysis was the hydrogenophile *Methanoculleus* (~10%) although it can also use formate (Maus *et al.*, 2012). This genus was absent in the thermophilic digesters and barely significant in the mesophilic digesters without thermal hydrolysis. Principal component analysis revealed that the thermophilic, mesophilic and mesophilic digesters with thermal hydrolysis had fundamentally different bacterial populations. Typically, digesters preceded by thermal hydrolysis had far less variability in bacterial populations. At the phylum level, both meso- and thermophilic digestion plants had large abundance (between <10% and 30% of each) of Firmicutes, Proteobacteria, Chloroflexi, Actinobacteria and Bacterioidetes. By contrast, digesters fed thermally hydrolyzed sludge were dominated by Firmicutes (<70%) with approximately 20% Bacterioidetes. Firmicutes are known to produce extracellular enzymes to enhance hydrolysis of cellulose, proteins, lignin and lipids, as well as use volatile fatty acids prior to acetic acid production. Therefore, the presence of Firmicutes is a measurement of improved hydrolysis (Chen *et al.*, 2018) and enhanced production of acetic acid for acetogenic methanogens.

This has been mirrored to an extent in a subsequent work which specifically looked at the difference between pre- and post-digestion thermal hydrolysis on the microbiome compared with mesophilic digestion as a control (Yang *et al.*, 2019). With the control digester Firmicutes comprised only 30% of counts but contributed 70% and approximately 50% for downstream and upstream hydrolysis respectively. Firmicutes produce extracellular enzymes which help with hydrolysis during anaerobic digestion, suggesting that digestion preceded by thermal hydrolysis has greater capacity for hydrolysis of soluble materials (Yang *et al.*, 2019). As with Kirkegaard's work from 2017, Firmicutes and Bacteroidetes combined were almost 90% of counts with pre-digestion hydrolysis. Post-digestion application shared similar microbiological traits with the mesophilic control with reasonable contributions of Actinobacteria and WS6, both of which were almost absent with the pre-digestion configuration. Synergistetes was significant in the post-thermal hydrolysis setup but was virtually absent in the other two configurations. It ferments acetic acid into the precursors of hydrogenotrophic methanogens in a syntrophic relationship (Jiang *et al.*, 2014, cited by Yang *et al.*, 2019).

As for the Danish study, quantification of archaea at the genus level was not dominated by a strict hydrogenotroph, but by the versatile *Methanosarcina* which can use a variety of substrates, which contributed approximately 70% of counts for mesophilic digestion with or without pre-digestion thermal hydrolysis, and this fell to 60% for post application (Yang *et al.*, 2019). One area where both hydrolysis options differed from the control was with levels of *Methanoculleus* – primarily a hydrogen consumer – which contributed between 10% and 20% counts for thermal hydrolysis in close agreement with Kirkegaard *et al.* (2017), but an order of magnitude less in the absence of thermal processing.

At the same time as Yang's work, a team was investigating the microbiome of two full-scale thermal hydrolysis plants as well as the fate of antibiotic-resistant genes in Beijing (Tong *et al.*, 2019). The raw sludge from both sites had large concentrations of *Methanosaeta* and *Methanospirillum*, and to a lesser extent *Methanosarcina*. However, in accordance with both the work of Kirkegaard and also Yang, thermal hydrolysis stimulated the growth of *Methanoculleus* in the digester, in spite of its absence in the raw sludge. It is clear that a better understanding of its kinetics is required to comprehend the impacts of thermal hydrolysis on digestion. *Methanoculleus* has a growth optima of 40°C (Asakawa & Nagoaka, 2003) which coincides with the operating temperature of most digestion plants following thermal hydrolysis.

In addition to that hydrogen utilizing organism, *Methanospirillum* was significant. Between them, they contributed approximately 70% of archaea counts within the full-scale digesters. Thermal hydrolysis was found to shift methanogenic populations from acetotrophs to hydrogenotrophs as would be expected due to higher FA. However, the feed sludge, unlike in other studies, was already enriched with hydrogen consumers. In the raw sludges, *Methanosaeta* and *Methanosarcina* together accounted for approximately 35% and 20% of counts at both plants respectively. After treatment, the digesters were measuring drops to 17% and 14%. Conversely, hydrogen consumers accounted for 50% and 60% of counts prior to treatment, and these figures increased to 61% and 67% during digestion (Tong *et al.*, 2019). This work was in agreement with that of Ennouri *et al.* (2016) who reported an increase in hydrogen-using methanogens measured by FISH as a consequence of thermal pre-treatment using an autoclave. With respect to bacteria, Tong's group showed that raw sludge bacteria were dominated by organisms which contribute to wastewater treatment, such as *Nitrospira* (nitrite-oxidizing bacteria), *Dechloromonas*, *Dokdonella* (both denitrifiers and phosphate accumulators), *Comanonas* and *Thauera* (versatile bacteria). After thermal hydrolysis, there was a dominance of *Clostridium sensu stricto*, *Clostridium XI*, *Kurthia* among others. These organisms are acid-tolerant, spore-forming bacteria with good thermostability and drying resistance (Wiegel *et al.*, 2006 cited by Tong *et al.*, 2019). After digestion, as with other studies, bacterial populations were dominated by Firmicutes at the phylum level (Tong *et al.*, 2019).

In Shanghai, Chen's team (2018) attempted to elucidate the impact of thermal hydrolysis (160°C, 30 min) on bacterial population with analysis of the different moieties within the sludge prior to and post treatment. As with other studies, data were collected over a time period to look for evolution within the digesters. Both digesters were fed at loading rates above 4.2 kg VS/m^3 d, and thermal hydrolysis improved volatile solids destruction from 39% to 52% corresponding to increased biogas yields from 0.35 to 0.51 m^3/kg VS_{fed}, in spite of TAN of 3.3 and 4.3 g/l in control and test reactors, respectively (Chen *et al.*, 2018). The authors found that improved degradation was down to better destruction of proteins (46% vs. 60%), lipids (30% vs. 51%), polysaccharides (49% vs. 71%) and hemicelluloses (20% vs. 48%). The research showed poor degradability of cellulose (between 5% and 9%) and lignin (<2% irrespective of pre-treatment, Chen *et al.*, 2018). Thermal hydrolysis released proteins and polysaccharides and lipids and therefore improved access for enzymes and micro-organisms. As with the aforementioned studies, the Firmicutes phylum dominated the digester combined with thermal hydrolysis, and this increased over time (70% of abundance after 65 d operation), largely at the expense of Bacteriodetes and Proteobacteria. *Clostridiales* accounted for approximately 85% of counts, in agreement with Tong's work (2019). *Clostridiales* has been identified as being able to follow the syntrophic acetate oxidation pathway, which results in the precursors of hydrogenotrophic methanogenesis. By contrast, Firmicutes reduced with time in the digester without thermal hydrolysis from 65% to 35% over the same time period being displaced by WS6. Furthermore, *Clostridiales* contributed a smaller percentage of the Firmicutes (circa 35%). Although significant difference was found with bacteria, samples for archaea were similar for both digesters, and coincide with the findings of Kirkegaard *et al.* (2017). In the Shanghai work, the predominant archaea were *Methanosarcina* for both control and test digesters. Furthermore, concentrations of *Methanosarcina* increased with time from 70% to 90%, and 85% to <100% relative abundance for digesters without and with thermal hydrolysis, respectively. A difference was also found in the levels of the hydrogenotroph *Methanobrevibacter* which was higher in the digester-fed hydrolyzed sludge. The workers concluded that methanogenic pathways had shifted from a strictly acetogenic one to one based on both acetogenic and hydrogenotrophic pathways. The study also looked at specific genes after 65 d of operation to determine how thermal hydrolysis was influencing digestion (Chen *et al.*, 2018). The genes responsible for transport and metabolism of amino acids and carbohydrates were enriched, as well as those used for energy production and conversion. However, the transport and metabolism of lipids fell in the test reactor. The improvement of volatile solids destruction by enhanced protein degradation was further studied by the same team (Chen *et al.*, 2019).

Ennouri *et al.* (2016) investigated thermal treatment at 60°C and also at 120°C to describe the main trophic groups during digestion using FISH. They found a large reduction in hydrolytic acidogens from 45% (in feed) to only 2% of cell population in digesters where sludge was heated to both 60°C and 120°C. As heating hydrolyses components in the sludge, it appears that the population of bacteria responsible for hydrolysis become redundant during digestion. This may explain the results of Li's group (2019) looking at combined thermal hydrolysis with recuperative thickening. In that work, digested sludge is recycled and no further improvement in performance is noted in spite of increased solid retention time. In addition to an increase in acetogenic bacteria, Ennouri's team (2016) also found increases in syntrophic propionate and butyrate specialists, which provide stability during digestion and precursors for methanogenesis. This was confirmed with increases in archaea from circa 45% of total cell number to between 75% and 85% after thermal manipulation. In accordance with previous work, both acetotrophic and hydrogenotrophic methanogen populations increased by 30% and nearly 300% for acetate and hydrogen utilizers, respectively, with 120°C preheating. Curiously, hydrogen utilizers increased by a factor of 4 with 60°C pre-treatment and then reduced at the higher temperature.

There are numerous other studies which study the microbiome when thermal hydrolysis is combined with digestion. But as described here, a number of trends are evident. Although a shift in methanogenic activity is noticed from acetogenic to hydrogenotrophic pathways, quantification is difficult, and digesters fed thermally hydrolyzed sludge may retain strong acetogenic pathways, with strong populations of *Methanosarcina*, which can use multiple substrates and more robust against ammonia toxicity. However, full-scale digesters have also shown *Methanosaeta* which do not partake in hydrogen-derived methanogenesis. Subsequently, digesters processing hydrolyzed sludge under mesophilic conditions share more traits in common with the microbiome of other mesophilic digesters rather than thermophilic counterparts where FA concentrations are higher resulting in a strong hydrogenotrophic response. An increase in methanogenic populations as a percentage of total organisms is noted after thermal hydrolysis which would explain higher methane yields. The species of *Methanoculleus* is promoted in digesters treating hydrolyzed material, and its abundance is fundamentally higher than in digesters where thermal hydrolysis is absent irrespective of digestion temperature or positioning of hydrolysis. Therefore, understanding of the kinetics and response of this species to environmental parameters is fundamental to further elucidate the impact of thermal hydrolysis on digestion. Following hydrolysis, digesters are dominated by Firmicutes at the phylum level. Firmicutes contain bacteria which specialize in the formation of methanogenic precursors and the breakdown of proteins and celluloses. An increasing ratio of Firmicutes to Bacteroidetes has been linked to a concomitant improvement in anaerobic digestion stability (Chen *et al.*, 2016) and this appears to be the case with thermal hydrolysis where these ratios have been measured over time. Interestingly, concentrations of hydrolytic acidogens are significantly diminished, due no doubt to the fact that the incoming feed material has been hydrolyzed, making their presence largely extraneous. This may mean that work to further improve hydrolysis rates during digestion of thermally hydrolyzed sludge may be limited in success (Li *et al.*, 2019) as the organisms to conduct the relevant reactions are absent. Positioning of thermal hydrolysis also has an influence of the microbiome of digestion. When positioned downstream, bacterial and archaeal populations are largely similar to standard digestion with the exception of *Methanoculleus* mentioned previously. Although, much research is still needed in this area, recent work is now shedding light onto the influence of thermal hydrolysis on microbiology within a digester thus opening options for further digestion enhancement.

REFERENCES

Abelleira J., Pérez-Elvira S. I., Portela J. R., Sánchez-Oneto J. and Nebot E. (2012). Advanced thermal hydrolysis: optimization of a novel thermochemical process to aid sewage sludge treatment. *Environmental Science & Technology*, **46**(11), 6158–6166.

Ahuja N. (2015). Impact of Operating Conditions on Thermal Hydrolysis Pre-treated Digestion Return Liquor. Thesis submitted to faculty of Virginia Polytechnic Institute and State University in partial fulfillment of the requirement for the degree of Masters of Science in Civil and Environmental Engineering.

Ajandouz E. H. and Puigserver A. (1999). Nonenzymatic browning reaction of essential amino acids: effect of pH on caramelization and Maillard reaction kinetics. *Journal of Agricultural and Food Chemistry*, **47**(5), 1786–1793.

Al-Shemmeri T. (2012). Engineering Fluid Mechanics. Ventus Publishing ApS, pp. 17–18. ISBN 978-87-403-0114-4.

Angelidaki I., Ellegaard L. and Ahring B.K. (1993). A mathematical model for dynamic simulation of anaerobic digestion of complex substrate: focusing on ammonia inhibition. *Biotechnology and Bioengineering*, **42**, 159–166.

Anson M. L. (1954). Advances in Protein Chemistry. Academic Press, New York, USA.

Anson M. and Mirsky A. (1932). The effect of denaturation on the viscosity of protein systems. *The Journal of General Physiology*, **15**(3), 341–350.

Aquino S. F. and Stuckey D. C. (2008). Integrated model of the production of soluble microbial products (SMP) and extracellular polymeric substances (EPS) in anaerobic chemostats during transient conditions. *Biochemical Engineering Journal*, **38**(2), 138–146.

Asakawa S. and Nagaoka K. (2003). Methanoculleus bourgensis, Methanoculleus olentangyi and Methanoculleus oldenburgensis are subjective synonyms. *International Journal of Systematic and Evolutionary Microbiology*, **53**(5), 1551–1552.

Baisier W. M. and Labuza T. P. (1992). Maillard browning kinetics in a liquid model system. *Journal of Agricultural and Food Chemistry*, **40**(5), 707–713.

Banks C. J., Zhang Y., Jiang Y. and Heaven S. (2012). Trace element requirements for stable food waste digestion at elevated ammonia concentrations. *Bioresource Technology*, **104**, 127–135.

Barber W. P. F., Peot C., Murthy S., Higgins M. and Bodniewicz B. (2015). The Potential for Co-digestion of Organics Using Thermal Hydrolysis at Blue Plains Advanced Wastewater Treatment Works. Proceedings of WEF Residuals and Biosolids 2015, Water and Environment Federation, Washington DC.

Barjenbruch M. and Kopplow O. (2003). Enzymatic, mechanical and thermal pre-treatment of surplus sludge. *Advances in Environmental Research*, **7**(3), 715–720.

Barker D. J. and Stuckey D. C. (1999). A review of soluble microbial products (SMP) in wastewater treatment systems. *Water Research*, **33**(14), 3063–3082.

Barnard J. and Shimp G. (2013). The Impacts of Biological Phosphorus Removal on Biosolids Dewatering. Proceedings of Joint IWA – WEF Conference on Nutrient Management, Vancouver.

Batstone D. (2017). Colour and Refractory COD Formation in THP. Presented at a Thermal Hydrolysis Workshop, Sludgetech Conference, 2017, London, England.

Baudez J. C., Slatter P. and Eshtiaghi N. (2013). The impact of temperature on the rheological behaviour of anaerobic digested sludge. *Chemical Engineering Journal*, **215**, 182–187.

Biochemistry of Foods. (2013). Chapter 6: Browning Reactions in Foods, pp. 245–289.

Borrelli R. C., Visconti A., Mennella C., Anese M. and Fogliano V. (2002). Chemical characterization and antioxidant properties of coffee melanoidins. *Journal of Agricultural and Food Chemistry*, **50**(22), 6527–6533.

Bougrier C., Albasi C., Delgenès J. P. and Carrère H. (2006). Effect of ultrasonic, thermal and ozone pre-treatments on waste activated sludge solubilisation and anaerobic biodegradability. *Chemical Engineering and Processing: Process Intensification*, **45** (8), 711–718.

Bougrier C., Delgenès J. P. and Carrère H. (2008). Effects of thermal treatments on five different waste activated sludge samples solubilisation, physical properties and anaerobic digestion. *Chemical Engineering Journal*, **139**(2), 236–244.

Braguglia G. M., Gianico A., Nakamura K. and Rossetti S. (2014). Thermophilic anaerobic digestion of thermal pretreated sludge: role of microbial community structure and correlation with process performances. Performances and microbial composition during mesophilic and thermophilic anaerobic sludge digestion processes, 75.

Bura R., Cheung M., Liao B., Finlayson J., Lee B. C., Droppo I. G., Leppard G. G. and Liss S. N. (1998). Composition of extracellular polymeric substances in the activated sludge floc matrix. *Water Science and Technology*, **37**(4–5), 325–333.

Chauzy J., Cretenot D., Bausseron A. and Deleris S. (2008). Anaerobic digestion enhanced by thermal hydrolysis: first reference BIOTHELYS® at Saumur. France. *Water Practice and Technology*, **3**(1), wpt2008004.

Chen Y., Cheng J. J. and Creamer K. S. (2008). Inhibition of anaerobic digestion process: a review. *Bioresource Technology*, **99**(10), 4044–4064.

Chen J. L., Ortiz R., Steele T. W. and Stuckey D. C. (2014). Toxicants inhibiting anaerobic digestion: a review. *Biotechnology Advances*, **32**(8), 1523–1534.

Chen S., Cheng H., Wyckoff K. N. and He Q. (2016). Linkages of Firmicutes and Bacteroidetes populations to methanogenic process performance. *Journal of Industrial Microbiology & Biotechnology*, **43**(6), 771–781.

Chen S., Ning L., Dong B., Zhao W., Dai L. and Dai X. (2018). New insights into the enhanced performance of high solid anaerobic digestion with dewatered sludge by thermal hydrolysis: organic matter degradation and methanogenic pathways. *Journal of Hazardous Materials*, **342**, 1–9.

Chen S., Dong B., Dai X., Wang H., Li N. and Yang D. (2019). Effects of thermal hydrolysis on the metabolism of amino acids in sewage sludge in anaerobic digestion. *Waste Management*, **88**, 309–318.

Christensen M. L., Keiding K., Nielsen P. H. and Jørgensen M. K. (2015). Dewatering in biological wastewater treatment: a review. *Water Research*, **82**, 14–24.

Chu C. P., Lin W. W., Lee D. J., Chang B. V. and Peng X. F. (2002). Thermal treatment of waste activated sludge using liquid boiling. *Journal of Environmental Engineering*, **128**(11), 1100–1103.

Dawson M. and Sotiriadis T. (2007). Effect of Modern Treatment Processes and Chemical Dosing on Sludge Rheology. Proceedings of Aquaenvio 12th European Biosolids and Organic Resources Conference and Exhibition, Manchester, UK.

Dawson M. K. and Ozgencil H. (2009). Do Hydrolysis Processes Effect Digested Sludge Rheology? Proceedings of Aquaenvio 14th European Biosolids and Organic Resources Conference and Exhibition, Leeds, UK.

deBarbadillo C. (2016). Integration of Thermal Hydrolysis, Anaerobic Digestion and Sidestream Deammonification at Blue Plains AWTP. Presented at IWA Singapore International Water Week.

Demirel B. and Scherer P. (2008). The roles of acetotrophic and hydrogenotrophic methanogens during anaerobic conversion of biomass to methane: a review. *Reviews in Environmental Science and Bio/Technology*, **7**(2), 173–190.

De Vrieze J., Saunders A. M., He Y., Fang J., Nielsen P. H., Verstraete W. and Boon N. (2015). Ammonia and temperature determine potential clustering in the anaerobic digestion microbiome. *Water Research*, **75**, 312–323.

Driessen W., van Veldhoven J. T. A., Hendrickx T. and Van Loosdrecht M. C. M. (2018). Successful Treatment of Sidestream Dewatering Liquors from Thermally Hydrolysed and Mesophilic Anaerobically Digested (THP-MAD) Biosolids. Proceedings of IWA Nutrient Removal and Recovery Conference, 18–21 November, Brisbane, Australia.

Dwyer J. and Lant P. (2008). Biodegradability of DOC and DON for UV/H2O2 pre-treated melanoidin based wastewater. *Biochemical Engineering Journal*, **42**(1), 47–54.

Dwyer J., Starrenburg D., Tait S., Barr K., Batstone D. J. and Lant P. (2008). Decreasing activated sludge thermal hydrolysis temperature reduces product colour, without decreasing degradability. *Water Research*, **42**(18), 4699–4709.

Dwyer J., Griffiths P. and Lant P. (2009). Simultaneous colour and DON removal from sewage treatment plant effluent: alum coagulation of melanoidin. *Water Research*, **43** (2), 553–561.

Echavarría A. P., Pagán J. and Ibarz A. (2012). Melanoidins formed by Maillard reaction in food and their biological activity. *Food Engineering Reviews*, **4**(4), 203–223.

Ennouri H., Miladi B., Diaz S. Z., Güelfo L. A. F., Solera R., Hamdi M. and Bouallagui H. (2016). Effect of thermal pretreatment on the biogas production and microbial communities balance during anaerobic digestion of urban and industrial waste activated sludge. *Bioresource Technology*, **214**, 184–191.

Eshtiaghi N., Markis F., Yap S. D., Baudez J. C. and Slatter P. (2013). Rheological characterisation of municipal sludge: a review. *Water Research*, **47**(15), 5493–5510.

Eskin M. N. A., Ho C.-T. and Shahidi F. (2013). Chapter 6: Browning Reactions in Food. In: Biochemistry of Foods, 3rd edn, Elsevier Publishing, London, pp. 245–281.

Everett J. G. (1972). Dewatering of wastewater sludge by heat treatment. *Journal Water Pollution Control Federation*, 92–100.

Everett J. G. (1974). The effect of pH on the heat treatment of sewage sludges. *Water Research*, **8**(11), 899–906.

Farno E., Baudez J. C., Parthasarathy R. and Eshtiaghi N. (2014). Rheological characterisation of thermally-treated anaerobic digested sludge: impact of temperature and thermal history. *Water Research*, **56**, 156–161.

Farno E., Baudez J. C., Parthasarathy R. and Eshtiaghi N. (2015). Impact of temperature and duration of thermal treatment on different concentrations of anaerobic digested sludge: kinetic similarity of organic matter solubilisation and sludge rheology. *Chemical Engineering Journal*, **273**, 534–542.

Fjordside C. (2005). Full Scale Experience of Retrofitting Thermal Hydrolysis to an Existing Anaerobic Digester for the Digestion of Waste Activated Sludge. Proceedings of the Aqua-Envio 16th Annual Residuals and Biosolids Management Conference, Manchester, UK.

Forster C. F. (1983). Bound water in sewage sludge and its relationship to sludge surfaces and sludge viscosities. *Journal of Chemical Technology and Biotechnology*, **33**(1), 76–84.

Fotidis I. A., Karakashev D., Kotsopoulos T. A., Martzopoulos G. G. and Angelidaki I. (2013). Effect of ammonium and acetate on methanogenic pathway and methanogenic community composition. *FEMS Microbiology Ecology*, **83**(1), 38–48.

Gagliano M. C. (2014). Performances and Microbial Composition during Mesophilic and Thermophilic Anaerobic Sludge Digestion Processes. PhD thesis, Università degli studi della Tuscia, Italy.

Geraats B. (2014). Lysotherm® sludge Hydrolysis. Five Year Experience with a Novel Approach for Operational Savings. Presented at: 19th European Biosolids & Organic Resources Conference & Exhibition, Manchester, UK.

Gerardi M. H. (2003). The Microbiology of Anaerobic Digesters. John Wile y& Sons, Inc., Hoboken, New Jersey.

Haile F., Ghoor T., Ikumi D., Wadhawan T., Brouckaert C. J., Murthy S. N., Al-Omari A. and Ekama G. (2015). A Stoichiometric and Elementally Balanced Model for Simulating Anaerobic Digester Failure. Proceedings of WEF Residuals and Biosolids 2015, Water and Environment Federation, Washington DC.

Hasan M. (2017). EPS Transformation through different Treatment Processes and its effect on Cake Solids. Presented at SludgeTech Conference, London, UK.

Haug R. T. (1977). Sludge processing to optimize digestibility and energy production. *Journal Water Pollution Control Federation*, 1713–1721.

Haug R. T., Stuckey D. C., Gossett J. M. and McCarty P. L. (1978). Effect of thermal pretreatment on digestibility and dewaterability of organic sludges. *Journal Water Pollution Control Federation*, 73–85.

Hendriks A. T. W. M. and Zeeman G. (2009). Pretreatments to enhance the digestibility of lignocellulosic biomass. *Bioresource Technology*, **100**(1), 10–18.

Higgins M. J. and Novak J. T. (1997). The effect of cations on the settling and dewatering of activated sludges: laboratory results. *Water Environment Research*, **69**, 215–224.

Higgins M., Beightol S., Mandahar U., Xiao S., Hung-Wei L., Le T., Mah J., Pathak B., Novak J., Al-Omari A. and Murthy S. (2015). Effect of Thermal Hydrolysis Temperature on Anaerobic Digestion, Dewatering and Filtrate Characteristics. Proceedings of WEFTEC 2015, New Orleans.

Higgins M. J., Beightol S., Mandahar U., Suzuki R., Xiao S., Lu H. W., Le T., Ma J., Pathak P., De Clippeleir H., Novak J. T., Al-Omari A. and Murthy S. N. (2017). Pretreatment of a primary and secondary sludge blend at different thermal hydrolysis temperatures: Impacts on anaerobic digestion, dewatering and filtrate characteristics. *Water Research*, **122**, 557–569.

Hii K., Parthasarathy R., Baroutian S., Gapes D. J. and Eshtiaghi N. (2017). Rheological measurements as a tool for monitoring the performance of high pressure and high temperature treatment of sewage sludge. *Water Research*, **114**, 254–263.

Hii K., Farno E., Baroutian S., Parthasarathy R. and Eshtiaghi N. (2019). Rheological characterization of thermal hydrolysed waste activated sludge. *Water Research*, **156**, 445–455.

Ho D. P., Jensen P. D. and Batstone D. J. (2013). Methanosarcinaceae and acetate-oxidizing pathways dominate in high-rate thermophilic anaerobic digestion of waste-activated sludge. *Applied and Environmental Microbiology*, **79**(20), 6491–6500.

Hodge J. E. (1953). Dehydrated foods, chemistry of browning reactions in model systems. *Journal of Agricultural and Food Chemistry*, **1**(15), 928–943.

Hodge J. E. (1955). The Amadori rearrangement. In: Advances in Carbohydrate Chemistry, Melville L. Wolfrom (ed.), Academic Press, London, vol. **10**, pp. 169–205.

Hung-Wei L., Xiao S., Le T., Al-Omari A., Higgins M., Boardman G., Novak J. and Murthy S. (2014). Evaluation of Solubilization Characteristics of Thermal Hydrolysis Process. Proceedings of WEFTEC.

Houghton J. J., Quarmby J. and Stephenson T. (2001). Municipal wastewater sludge dewaterability and the presence of microbial extracellular polymer. *Water Science and Technology*, **44**(2–3), 373–379.

Jiang J., Wu J., Poncin S. and Li H. Z. (2014). Rheological characteristics of highly concentrated anaerobic digested sludge. *Biochemical Engineering Journal*, **86**, 57–61.

Kato H. and Tsuchida H. (1981). Estimation of melanoidin structure by pyrolysis and oxidation. *Progress In Food and Nutrition Science*, **5**, 147–156.

Kaushik A., Basu S., Singh K., Batra V. S. and Balakrishnan M. (2017). Activated carbon from sugarcane bagasse ash for melanoidins recovery. *Journal of Environmental Management*, **200**, 29–34.

Kayhanian M. (1999). Ammonia inhibition in high-solids biogasification: an overview and practical solutions. *Environmental Technology*, **20**, 355–365.

Kepp U. and Solheim O. E. (2000). Thermo Dynamical Assessment of the Digestion Process. In CIWEM/Aqua Enviro 5th European Biosolids and Organic Residiuals Conference.

Kirkegaard R. H., McIlroy S. J., Kristensen J. M., Nierychlo M., Karst S. M., Dueholm M. S.,... and Nielsen P. H. (2017). The impact of immigration on microbial community composition in full-scale anaerobic digesters. *Scientific Reports*, **7**(1), 9343.

Kitaoka S. and Suzuki K. (1967). Caramels and caramelization. *Agricultural and Biological Chemistry*, **31**(6), 753–755.

Kuhn R. and Weygand F. (1937). Die Amadori-Umlagerung. *Berichte der deutschen chemischen Gesellschaft (A and B Series)*, **70**(4), 769–772.

Kumar P. and Chandra R. (2006). Decolourisation and detoxification of synthetic molasses melanoidins by individual and mixed cultures of *Bacillus* spp. *Bioresource Technology*, **97**(16), 2096–2102.

Kus F. and Wiesmann U. (1995). Degradation kinetics of acetate and propionate by immobilized anaerobic mixed cultures. *Water Research*, **29**(6), 1437–1443.

Labuza T. and Baisier W. M. (1992). The kinetics of nonenzymatic browning. In: Physical Chemistry of Foods, Marcel Dekker, New York, pp. 595–649.

Lancaster R. (2015). Thermal Hydrolysis at Davyhulme WWtW – One Year On. Proceedings of WEF Residuals and Biosolids 2015, Water and Environment Federation, Washington DC.

Lay J. J., Li Y. Y. and Noike T. (1998). The influence of pH and ammonia concentration on the methane production in high-solids digestion processes. *Water Environment Research*, **70**(5), 1075–1082.

Li Y. Y. and Noike T. (1992). Upgrading of anaerobic digestion of waste activated sludge by thermal pretreatment. *Water Science and Technology*, **26**(3–4), 857–866.

Li B., Romero A., Wadhawan T., Tobin M., Manning E., Higgins M., Al-Omari A., Murthy S., Riffat R. and De Clippeleir H. (2019). Recuperative thickening for sludge retention time and throughput management in anaerobic digestion with thermal hydrolysis pretreatment. *Water Environment Research*, **1**, 1–13.

Liu T. and Sung S. (2002). Ammonia inhibition on thermophilic aceticlastic methanogens. *Water Science and Technology*, **45**(10), 113–120.

Liu C. S., Higgins C. P., Wang F. and Shih K. (2012a). Effect of temperature on oxidative transformation of perfluorooctanoic acid (PFOA) by persulfate activation in water. *Separation and Purification Technology*, **91**, 46–51.

Liu X., Wang W., Gao X., Zhou Y. and Shen R. (2012b). Effect of thermal pretreatment on the physical and chemical properties of municipal biomass waste. *Waste Management*, **32**(2), 249–255.

Lotito V. and Lotito A. M. (2014). Rheological measurements on different types of sewage sludge for pumping design. *Journal of Environmental Management*, **137**, 189–196.

Lu F., Zhou Q., Wu D., Wang T. F., Shao L. M. and He P. J. (2015). Dewaterability of anaerobic digestate from food waste: relationship with extracellular polymeric substances. *Chemical Engineering Journal*, **262**, 932–938.

Lumb C. (1940). Heat treatment as an aid to sludge dewatering-ten months" full-scale operation. *Surveyor*, **98**(2550), 287–91.

Lumb C. (1951). Heat treatment as an aid to sludge dewatering: ten years" full-scale operation. *Journal of the Proceedings of the Institute of Sewage Purification*, No. 1, 5.

Maillard L. C. (1912). Action des acides amines sur les sucres; formation des melanoidines par voie methodique. *Comptes Rendus de l'Académie des Sciences (Paris)*, **154**, 66–68.

Markis F., Baudez J. C., Parthasarathy R., Slatter P. and Eshtiaghi N. (2014). Rheological characterisation of primary and secondary sludge: impact of solids concentration. *Chemical Engineering Journal*, **253**, 526–537.

Markis F., Baudez J. C., Parthasarathy R., Slatter P. and Eshtiaghi N. (2016). The apparent viscosity and yield stress of mixtures of primary and secondary sludge: impact of volume fraction of secondary sludge and total solids concentration. *Chemical Engineering Journal*, **288**, 577–587.

Martins S. I., Jongen W. M. and Van Boekel M. A. (2000). A review of Maillard reaction in food and implications to kinetic modelling. *Trends in Food Science & Technology*, **11**(9–10), 364–373.

Maus I., Wibberg D., Stantscheff R., Eikmeyer F. G., Seffner A., Boelter J.,... and Pühler A. (2012). Complete genome sequence of the hydrogenotrophic, methanogenic archaeon Methanoculleus bourgensis strain MS2T, isolated from a sewage sludge digester. *Journal of Bacteriology*, **194**, 5487–5488.

McCarty P. L. (1971). Energetics and kinetics of anaerobic treatment – Chapter 6. In: Anaerobic Biological Treatment Processes, R. F. Gould (ed.) 159th Meeting of the American Chemical Society, Houston, Tex., Advances in Chemistry Series No 105, Washington D.C.

McCarty P. L. and McKinney R. E. (1961). Salt toxicity in anaerobic digestion. *Journal Water Pollution Control Federation*, 399–415.

Merry J. and Oliver B. (2015). A Comparison of Real AD Plant Performance: Howdon, Bran Sands, Cardiff and Afan. In Proceedings of Aquaenviro's 20th European Biosolids and Organic Resources Conference and Exhibition, Manchester, UK.

Metcalf and Eddy. (2013). 5th edn, Wastewater Engineering: Treatment and Reuse. McGraw-Hill, New York, NY.

Migo V. P., Matsumura M., Del Rosario E. J. and Kataoka H. (1993). Decolorization of molasses wastewater using an inorganic flocculant. *Journal of Fermentation and Bioengineering*, **75**, 438–442.

Migo V. P., Del Rosario E. J. and Matsumura M. (1997). Flocculation of melanoidins induced by inorganic ions. *Journal of Fermentation and Bioengineering*, **83**(3), 287–291.

Mills N., Martinicca H., Fountain P., Shana A., Ouki S. and Thorpe R. (2013). Second Generation Thermal Hydrolysis Process. 18th European Biosolids and Biowastes Conference, Manchester, UK.

Molokwu O. and Rus E. (2017). Interactions of cations, phosphates, M/D and EPS. Presented at SludgeTech Conference, London, UK.

Monte W. C. and Maga J. A. (1981). Flavor chemistry of sucrose. *Sugar Technology Reviews*, **8**, 181–204.

Morales F. J. (2002). Application of capillary zone electrophoresis to the study of food and food-model melanoidins. *Food Chemistry*, **76**(3), 363–369.

Morales F. J., Fernández-Fraguas C. and Jiménez-Pérez S. (2005). Iron-binding ability of melanoidins from food and model systems. *Food Chemistry*, **90**(4), 821–827.

Murthy S. N. and Novak J. T. (1998). Effect of potassium ion on sludge settling, dewatering and effluent quality. *Water Science and Technology*, **37**, 317–324.

Nafisi K. and Markakis P. (1983). Inhibition of sugareamine browning by aspartic and glutamic acids. *Journal of Agricultural and Food Chemistry*, **31**, 1115–1117.

Nawawi D., Chauzy J., Gilbert A. B. and Cretenot D. (2007). Feed back of the operation of thermal pre-treatment on waste activated sludge before mesophilic anaerobic digestion. In: 12th European Biosolids and Organic Resources Conference, November 2007, P. Lowe and N.J. Horan (eds.), Aqua Enviro, Manchester, UK.

Neyens E. and Baeyens J. (2003). A review of thermal sludge pre-treatment processes to improve dewaterability. *Journal of Hazardous Materials*, **98**(1), 51–67.

Neyens E., Baeyens J. and Dewil R. (2004). Advanced sludge treatment affects extracellular polymeric substances to improve activated sludge dewatering. *Journal of Hazardous Materials*, **106**(2), 83–92.

Nguyen T., Fan L. and Roddick F. A. (2010). Removal of melanoidins from an industrial wastewater. In: Ozwater 10, edited by Australian Water Association-AWA, Sydney, NSW, Australia.

Ngwenya Z., Beightol S., Ngone Oo T., Vega J., Pathak B., Al-Omari A., Zhu K., Wadhawan T., Murthy S. N. and Higgins M. J. (2015). A Stoichiometric A Roach to Control Digester Chemistry and Ammonia Inhibition in Anaerobic Digestion with Thermal Hydrolysis Pretreatment: Model Development. Proceedings of WEF Residuals and Biosolids 2015, Water and Environment Federation, Washington DC.

Nursten H. (2005). The Maillard Reaction: Chemistry, Biochemistry and Implications. Royal Society of Chemistry, London.

Oerke D., Higgins M., Murthy S., Kharkar S., Debarbadillo C., Chiu C. and Rizzi S. (2017). State-of the-practice for dewatering of thermally hydrolyzed digested biosolids. *Proceedings of the Water Environment Federation Residuals and Biosolids Conference*, **2017**(6), 4471–4490.

Ohmomo S., Daengsubha W., Yoshikawa H., Yui M., Nozaki K., Nakajima T. and Nakamura I. (1988). Screening of anaerobic bacteria with the ability to decolorize molasses melanoidin. *Agricultural and Biological Chemistry*, **52**(10), 2429–2435.

O'Melia C. R. (1972). Coagulation and flocculation. In: Physico-Chemical Process for Water Quality Control. W. J. Weber, Jr. (ed.), Wiley and Sons Inc., New York, pp. 61–109.

Oosterhuis M., Ringoot D., Hendriks A. and Roeleveld P. (2014). Thermal hydrolysis of waste activated sludge at Hengelo wastewater treatment plant, The Netherlands. *Water Science and Technology*, **70**(1), 1–7.

Oracz J. and Nebesny E. (2019). Effect of roasting parameters on the physicochemical characteristics of high-molecular-weight Maillard reaction products isolated from cocoa beans of different *Theobroma cacao* L. *groups. European Food Research and Technology*, **245**(1), 111–128.

Pace G., Khunjar W., Yi P., Ishii S., Dole E., Santos M., ... and Joshi R. (2018). Impact of ozonation and electrocoagulation on refractory nutrients generated from thermal hydrolysis processes. *Proceedings of the Water Environment Federation*, **2018**(14), 2105–2107.

Panter K. (2009). Comparison of centrifuge and belt press for compressible digested sludge after thermal hydrolysis. *Proceedings of the Water Environment Federation*, **2009**(9), 6220–6229.

Pei J., Yao H., Wang H., Ren J. and Yu X. (2016). Comparison of ozone and thermal hydrolysis combined with anaerobic digestion for municipal and pharmaceutical waste sludge with tetracycline resistance genes. *Water Research*, **99**, 122–128.

Pereboom J., Luning L., Hol A., van Dijk L. and de Man A. W. A. (2014). Full Scale Experiences with TurboTec® Continuous Thermal Hydrolysis at WWTP Venlo (NL) and Apeldoorn (NL). Proceedings of Aqua-Enviro 19th European Biosolids and Organic Residuals Conference and Exhibition, Manchester, UK.

Pevere A., Guibaud G., Van Hullebusch E., Lens P. and Baudu M. (2006). Viscosity evolution of anaerobic granular sludge. *Biochemical Engineering Journal*, **27**(3), 315–322.

Phothilangka P., Schoen M. A. and Wett B. (2008). Benefits and drawbacks of thermal pre-hydrolysis for operational performance of wastewater treatment plants. *Water Science and Technology*, **58**(8), 1547.

Qiao W., Yin Z., Wang W., Wang J. and Zhang Z. (2013). Pilot-scale experiment on thermally hydrolyzed sludge liquor anaerobic digestion using a mesophilic expanded granular sludge bed reactor. *Water Science & Technology*, **68**(4).

Riches S., Irwin R. and Crassweller J. (2008). Integration of Thermal Hydrolysis into Sludge Treatment at Cottonvalley WwTW, Milton Keynes. Proceedings of the 13th Aquaenviro biosolids conference.

Rus E., Perrault A., Mills N., Shana A., Molokwu O. and Nilsen P. J. (2016). Optimizing THP – The Intermediate Thermal Hydrolysis Process. Proceedings of WEF Residuals and Biosolids Conference, Milwaukee, WI, USA.

Satyawali Y. and Balakrishnan M. (2008). Wastewater treatment in molasses-based alcohol distilleries for COD and color removal: a review. *Journal of Environmental Management*, **86**(3), 481–497.

Schnürer A. and Nordberg Å. (2008). Ammonia, a selective agent for methane production by syntrophic acetate oxidation at mesophilic temperature. *Water Science and Technology*, **57**(5), 735–740.

Sheerwood R. and Phillips J. (1970). Heat treatment process improves economics of sludge handling and disposal. *Water and Wastes Engineering*, Nov. 1970.

Simaratanamongkol A. and Thiravetyan P. (2010). Decolorization of melanoidin by activated carbon obtained from bagasse bottom ash. *Journal of Food Engineering*, **96**(1), 14–17.

Skinner S. J., Studer L. J., Dixon D. R., Hillis P., Rees C. A., Wall R. C. and Scales P. J. (2015). Quantification of wastewater sludge dewatering. *Water Research*, **82**(1), 2–13.

Sprott G. D. and Patel G. B. (1986). Ammonia toxicity in pure cultures of methanogenic bacteria. *Systematic and Applied Microbiology*, **7**(2–3), 358–363.

Sprott D. G., Shaw K. M. and Jarrel K. F. (1984). Ammonia/potassium exchange in methanogenic bacteria. *The Journal of Biological Chemistry*, **259**(20), 12602–12608.

Stickland A. D. (2015). Compressional rheology: a tool for understanding compressibility effects in sludge dewatering. *Water Research*, **82**, 37–46.

Stuckey D. C. and McCarty P. L. (1984). The effect of thermal pretreatment on the anaerobic biodegradability and toxicity of waste activated sludge. *Water Research*, **18**(11), 1343–1353.

Subramanian S. (2004). Digestion processes and their effect on dewatering and Bound water content of sludge. Civil and Environmental Engineering. Blacksburg, VA, Virginia Polytechnic Institute and State University, 92.

Svennevik O. K., Beck G., Rus E., Westereng B., Higgins M., Solheim O. E. and Horn S. J. (2019a). CNash-A novel parameter predicting cake solids of dewatered digestates. *Water Research*, **158**, 350–358.

Svennevik O. K., Solheim O. E., Beck G., Sørland G. H., Jonassen K. R., Rus E., Westereng B., Horn S. J., Higgins M. J. and Nilsen P. J. (2019b). Effects of post anaerobic digestion thermal hydrolysis on dewaterability and moisture distribution in digestates. *Water Science and Technology*, **80**(7), 1338–1346.

Taboada-Santos A., Braz G. H., Fernandez-Gonzalez N., Carballa M. and Lema J. M. (2019). Thermal hydrolysis of sewage sludge partially removes organic micropollutants but does not enhance their anaerobic biotransformation. *Science of the Total Environment*, **690**, 534–542.

Tampio E., Ervasti S., Paavola T., Heaven S., Banks C. and Rintala J. (2014). Anaerobic digestion of autoclaved and untreated food waste. *Waste Management*, **34**(2), 370–377.

Tian Y. and Zheng L. (2006). Functions and behaviors of activated sludge extracellular polymeric substances (EPS): a promising environmental interest. *Journal of Environmental Sciences*, **18**(3), 420–427.

Tong X., McCarty P. L. and Isaacson R. (1991). Microbial hydrolysis of lignocellulosic materials. *Methane from Community Wastes*, 61–100.

Tong J., Fang P., Zhang J., Wei Y., Su Y. and Zhang Y. (2019). Microbial community evolution and fate of antibiotic resistance genes during sludge treatment in two full-scale anaerobic digestion plants with thermal hydrolysis pretreatment. *Bioresource Technology*, 121575.

Urrea J. L., Collado S., Laca A. and Díaz M. (2015). Rheological behaviour of activated sludge treated by thermal hydrolysis. *Journal of Water Process Engineering*, **5**, 153–159.

Vesilind P. A. and Hsu C. C. (1997). Limits of sludge dewaterability. *Water Science and Technology*, **36**(11), 87–91.

Walley P. (2007). Optimising thermal hydrolysis for reliable high digester solids nloading and performance. In: 12th European Biosolids and Organic Resources Conference, 2007, P. Lowe and N.J. Horan (eds), Aqua Enviro, Manchester, UK.

Wang H., Fotidis I. A. and Angelidaki I. (2015). Ammonia effect on hydrogenotrophic methanogens and syntrophic acetate oxidizing bacteria. *FEMS Microbiology Ecology*, fiv130.

Wang J.-W., Li W., Jian Q.-H, Han X.-Y., Hao S. and Tang C.-C. (2018). Digestion Enhancement with Thermal Hydrolysis Pretreatment at Beijing's Gaobeidian Water Reclamation Plant – Operating Experience and Challenges. Proceedings of Water Environment Federation WEFTEC conference, pp. 2821–2834.

Wang L., Chang Y. and Li A. (2019). Hydrothermal carbonization for energy-efficient processing of sewage sludge: a review. *Renewable and Sustainable Energy Reviews*, **108**, 423–440.

Ward A. J., Arola K., Brewster E. T., Mehta C. M. and Batstone D. J. (2018). Nutrient recovery from wastewater through pilot scale electrodialysis. *Water Research*, **135**, 57–65.

Wett B., Takács I., Batstone D., Wilson C. and Murthy S. (2014). Anaerobic model for high-solids or high-temperature digestion-additional pathway of acetate oxidation. *Water Science and Technology*, **69**(8).

Wiegant W. M. and Zeeman G. (1986). The mechanism of ammonia inhibition in the thermophilic digestion of livestock wastes. *Agricultural Wastes*, **16**(4), 243–253.

Wiegel J., Tanner R. and Rainey F. A. (2006). An introduction to the family Clostridiaceae. In: The Prokaryotes, M. Dworkin, S. Falkow, E. Rosenberg, K.H. Schleifer and E. Stackebrandt (eds), Springer, New York, NY.

Wilson C. A. and Novak J. T. (2009). Hydrolysis of macromolecular components of primary and secondary wastewater sludge by thermal hydrolytic pretreatment. *Water Research*, **43**(18), 4489–4498.

Wilson C. A., Murthy S. M., Fang Y. and Novak J. T. (2008). The effect of temperature on the performance and stability of thermophilic anaerobic digestion. *Water Science and Technology*, **57**(2), 297.

Wilson C. A., Novak J., Takacs I., Wett B. and Murthy S. (2012). The kinetics of process dependent ammonia inhibition of methanogenesis from acetic acid. *Water Research*, **46**(19), 6247–6256.

Wilson C. A., Tanneru C. T., Banjade S., Murthy S. N. and Novak J. T. (2011). Anaerobic digestion of raw and thermally hydrolyzed wastewater solids under various operational conditions. *Water Environment Research*, **83**(9), 815–825.

Xue Y., Liu H., Chen S., Dichtl N., Dai X. and Li N. (2015). Effects of thermal hydrolysis on organic matter solubilization and anaerobic digestion of high solid sludge. *Chemical Engineering Journal*, **264**, 174–180.

Yang D., Hu C., Dai L., Liu Z., Dong B. and Dai X. (2019). Post-thermal hydrolysis and centrate recirculation for enhancing anaerobic digestion of sewage sludge. *Waste Management*, **92**, 39–48.

Yenigün O. and Demirel B. (2013). Ammonia inhibition in anaerobic digestion: a review. *Process Biochemistry*, **48**(5–6), 901–911.

Zhang C., Yuan Q. and Lu Y. (2014). Inhibitory effects of ammonia on methanogen mcrA transcripts in anaerobic digester sludge. *FEMS Microbiology Ecology*, **87**(2), 368–377.

Zhang J., Xue Y., Eshtiaghi N., Dai X., Tao W. and Li Z. (2017). Evaluation of thermal hydrolysis efficiency of mechanically dewatered sewage sludge via rheological measurement. *Water Research*, **116**, 34–43.

Zhang Q., Vlaeminck S. E., DeBarbadillo C., Su C., Al-Omari A., Wett B.,... and De Clippeleir H. (2018). Supernatant organics from anaerobic digestion after thermal hydrolysis cause direct and/or diffusional activity loss for nitritation and anammox. *Water Research*, **143**, 270–281.

Chapter 4

Operational experience

This chapter summarizes the shared experiences of the field application of thermal hydrolysis, especially with respect to starting digesters, operational knowhow, production and treatment of return liquors from dewatering digestate and co-digestion experience.

4.1 START-UP OF ANAEROBIC DIGESTION WITH THERMAL HYDROLYSIS

Successful start-up of an anaerobic digester involves the development of high and stable removal efficiencies in the shortest possible time, in order to maximize the financial and environmental benefits of its installation. During start-up, various trophic groups may not be in balance, and the reactor could fail; hence development of an optimum start-up strategy is extremely important. Several factors are important in the start-up of high rate systems (Killilea *et al.* 2000; Stronach *et al.*, 1986; Hickey *et al.*, 1991; Weiland & Rozzi, 1991) and these include volume and type of inoculum, wastewater, composition and strength and fluctuations in these parameters, biomass activity, growth rates, saturation constants, yield, adaption, ability to excrete polysaccharides, size and properties of granules, reactor configuration, geometry, size and ability to immobilize biomass, loading rate, hydraulic residence time (HRT) and mixing characteristics, environmental parameters such as pH and temperature and finally, the availability of nutrients and trace elements. Hence, developing an optimum start-up strategy can be complex and time consuming. During start-up, fluctuations in external

Author: William Barber
doi: 10.2166/9781789060287_0101

parameters such as temperature, pH, HRT and recycle ratio must be avoided and organic loading rates should also be consistent. Typically, initial loading rates should be low, with values of circa 0.1 kg COD/kg $VS_{in\ digester}$/d being reported in the literature for standard anaerobic treatment with no pre-treatment of the feedstock (Stronach *et al.*, 1986). For typical mixed sludge with a COD:VS ratio of 1.5, this is equivalent to approximately 0.067 kg VS_{fed}/kg $VS_{in\ digester}$/d. This value needs to be compared with typical loading rates when the digesters are at design loading. This is shown graphically in Figure 4.1.

For standard digestion, feeding rates of volatile solids can vary between 0.09 and 0.12 kg VS_{fed}/kg $VS_{in\ digester}$/d, compared with 0.18–0.25 kg VS_{fed}/kg $VS_{in\ digester}$/d of digester contents when thermal hydrolysis is present. These numbers are consistent with the work done by Thames Water (Fountain *et al.*, 2015). Therefore, these values are the target loading for stable operation. The time required for start-up depends critically on the quantity and type of seed inoculum available. Preferably, as much as possible similar seed, i.e., taken from an existing digester which is fed thermally hydrolyzed sludge, should be used. However, in many instances, this may be neither possible nor practical. For treating 100 t dry solids (DSs) per d with a digestion retention time of 14 d, for example, would require a seed volume of 14,000 m^3 to fill the digesters. Based on typical truck sizes this volume would require over 600 traffic movements into the site. Although not initially obvious, waste-activated sludge makes a good inoculum. Not only it is abundant in methanogens (Speece, 2008), but because of higher nitrogen concentration it has elevated quantities of acetate-oxidizing organisms compared with primary sludge. This provides more of the required bacteria for stable operation and releases greater concentrations of ammonia which increases pH. Furthermore, it has large concentrations of facultative anaerobes. It is important to include primary sludge in the seed as well due to concentrations of methanogens (Gerardi, 2003).

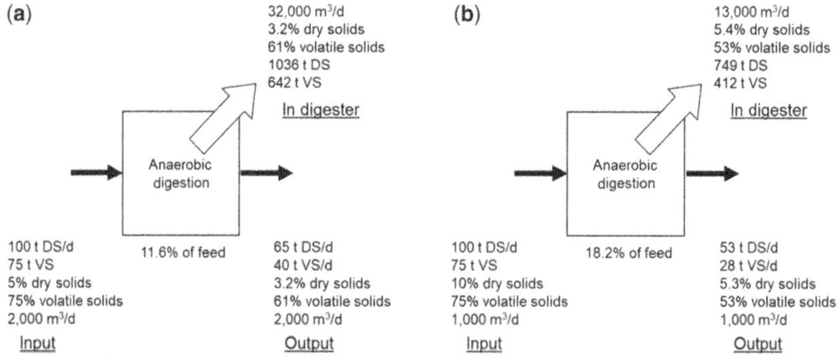

Figure 4.1 Comparison of digester inputs and outputs to digester contents: (a) standard digestion and (b) digestion preceded by thermal hydrolysis. Based on 100 t DSs fed/d.

Based on full-scale experience, it can take anywhere between 2 and 18 weeks (1–6 HRT at 20 d) to reach design throughput with a pseudo-steady-state operation, although some sites have taken longer (Belshaw *et al.*, 2013).

Typically, it takes 6 HRT to reach pseudo-steady state for starting up new digesters with thermal hydrolysis once they have been filled.

Figure 4.2 shows the approximate number of days required to reach steady-state, defined as having the full amount of volatile solids within the digester based on design loading, depending on quantity and type of seed added and also daily addition rate of feed.

Figure 4.2 shows that having thermally hydrolyzed sludge as an inoculum can almost eliminate the time required for start-up entirely if added at large quantity. In an enlightening study by Thames Water (Fountain *et al.*, 2015) based on long-term ownership of multiple facilities, the authors suggest that a grace period of approximately 2 weeks be added at the start-up of a digestion plant, regardless of seed type used. This is to allow microorganisms to acclimate to their new environment. This would add 14 d to the numbers proposed in Figure 4.2a and b. The figures suggest that if large quantities of non thermally hydrolyzed digested inoculum are used, it is necessary to increase the daily ramp up to reduce overall start-up time.

Some plants have been started up by bringing dewatered thermally hydrolyzed digested cake and then rewetting to the appropriate DSs for digestion to reduce traffic movements. If rewetted, it is necessary to reintroduce alkalinity in the dilution water to replace what was lost to the centrate from dewatering. At Davyhulme, 1 dry t sodium bicarbonate was added per 20 wet t of thermally

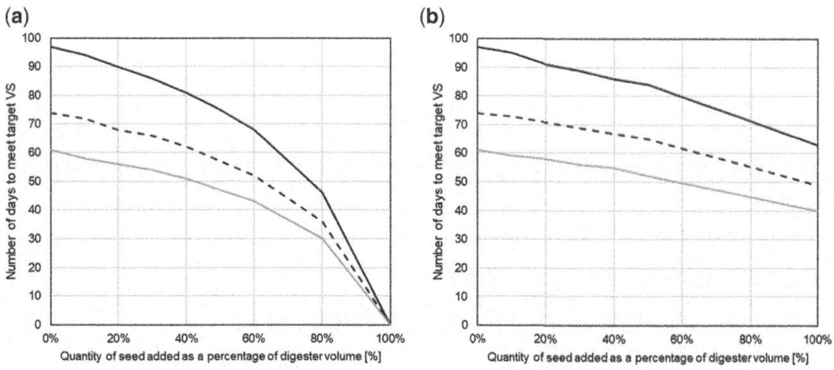

Figure 4.2 Time required to meet target VS quantity in digestion plant during start-up based on quantity of seed added and type of seed: (a) seed from thermal hydrolysis plant, DS% 5.5, VS% 55 and (b) seed from mesophilic digestion plant DS% 3.2, VS% 62%. *Key*: Black line = 3% daily addition of feed, black dashed line = 4%, gray line = 5%.

hydrolyzed digested cake to give an alkalinity of 6,000 mg/l $CaCO_3$ (Tattersall *et al.*, 2014).

Alkalinity requirements can be determined from the following equation (Speece, 2008):

$$HCO_3^- = \frac{K_1 K_H P_{CO_2}}{[H^+]} \tag{4.1}$$

where HCO_3^- = alkalinity (moles); $K_1 = 5.04 \times 10^{-7}$ and $K_H = 0.233 \times 10^{-1}$ (atm/mol frac) at 40°C.

Question. Example: how much alkalinity is required if using sodium bicarbonate for dilution water for rewetting thermally hydrolyzed digested cake from 30% DS to 5.5% DS, to ensure a pH of 7.6 with 35% carbon dioxide in biogas in a digester operating at 40°C?

Alkalinity required

$= (5.04 \times 10^{-7} \times 0.233 \times 10^{-1} \times 0.35)/10^{-7.6}$
$= 0.1636\ M \times 61\ g/1\ per\ M \times 1{,}000\ mg/1$

$= 9{,}981\ mg/1$

For 1 dry t of inoculum, total volume

$= 1/30\%$
$= 3.33$ wet t

Rewetted volume

$= 1/5.5\%$

$= 18.18$ wet t

Therefore, water for rewetting

$= 18.18$ wet t $- 3.33$ wet t $= 14.85$ wet t

For a concentration of 9,981 mg/l, a quantity of 148 kg $CaCO_3$ is required per 1 t DSs of inoculum.

Using sodium bicarbonate $NaHCO_3$

$CaCO_3 + H_2O + CO_2 \rightarrow Ca(HCO_3)_2$

$CaCO_3$ has a molecular weight of 100, and HCO_3 of 61

100 g $CaCO_3$ is equivalent to $(2 \times 61 =)$ 122 of HCO_3 (ratio of 1.22)

Therefore, quantity of $NaHCO_3$ required $= 1.22 \times 148$ kg $= 180$ kg/t DS inoculum

As with sodium bicarbonate, other additives have been used to aid start-up of anaerobic digesters. These include: caustic soda (NaOH), quicklime (CaO) and hydrated ($Ca(OH)_2$) lime and soda ash (Na_2CO_3) (Gerardi, 2003).

During start-up, there have been reported instances where the digesters become partially unstable. This has been hypothesized to be due to ammonia levels reaching a point where methanogenic pathways are altered from strictly acetotrophic (scavenging Methanosaetaceae and also Methanosarcinaceae) to a combination of acetate oxidation and hydrogenotrophic (mainly Methanosarcinaceae and smaller populations of hydrogenotrophs from the Methanomicrobiaceae family). This level is in excess of 1,500 mg/l total ammonia nitrogen (TAN) (Gerardi, 2003). This phenomenon has been studied previously. Speece and Parkin (1983) noted that the addition of 10,000 mg/l TAN at neutral pH caused biogas production to fall to zero. However, after 10 d of negligible performance, biogas levels rose back to over 70% of what they were within a further 5 d. The work showed that methanogens can acclimate to high ammonia levels given adequate time.

On sites with multiple digesters, it is common to seed digesters in pairs, with the first pair receiving inoculum from another site as discussed, and then when at pseudo-steady-state, those digesters seed subsequent pairs of digesters. As the plant progress, each pair takes progressively less time to reach design values. At Davyhulme, four pairs of digesters took 70, 55, 34 and 25 d respectively to reach design loadings (Belshaw *et al.*, 2013), approximately eight retention times. The first pair took longer to reach steady-state due to issues with maintaining consistent feed solids (Jolly *et al.*, 2012). The following steps are proposed for digester start-up based on full-scale experiences:

(1) Prepare digesters for start-up
  (a) New sites: wet and dry commissioning of digester and auxiliary equipment
  (b) Pre-existing: clean digesters, check mixing systems are adequate, nitrogen/CO_2 purging of headspace, check digester integrity
(2) Obtain seed inoculum, preferably from the THP plant. This may take several weeks depending on quantity of seed supplemented
(3) Allow up to 2 weeks for inoculum to acclimate *in situ*
(4) Ensure ancillary equipment and utilities (distribution systems; heat exchangers; water; steam; electrical power; compressed air; PLC; drainage systems; biogas infrastructure; instrumentation etc.) are ready and commissioned
(5) Monitor key variables which give indication of digester health: pH, alkalinity, total and volatile solids concentration, volatile fatty acids concentration, biogas production rate and composition
(6) Ramp up slowly, target is 0.18–0.25 kg VS_{fed}/kg $VS_{in\ digester}$/d, typically 3–5% increase daily
  (a) Speed of ramp up is more significant than quantity of seed
  (b) Alkalinity can be used to increase the rate of start-up
(7) Slow down feeding when TAN approaches the range of 1,500–2,000 mg/l
(8) Ramp up further to design throughput when digester performance stabilizes

4.1.1 Site experience

From site experience, it is critical to keep consistent feed quality and quantity and digester temperature consistent throughout both the start-up and operating phase of the thermal hydrolysis and digestion plants.

As with all new facilities, there are always several unforeseeable events which influence the time required for start-up. Generally, these include equipment not being up to specification, construction debris and foreign objects in equipment or sludge, spills and leaks, weather-related issues, incorrect time sequence of installation, limitations of certain plant equipment, failure of ancillary equipment, faulty valves or instrumentation and inconsistent feeding and composition of sludge. Table 4.1 highlights some of the published findings on the subject.

Table 4.1 Observations during start-up and commissioning of thermal hydrolysis plants with digestion.

Comments	Reference
• Steam not available for hydrolysis. Required back-up temporary boiler system • Pre-dewatering delayed • Extreme cold weather events • Steam used to accelerate digester heating, but diluted feed • Need to keep digesters virtually full to enable use of draft tube mixing	Loomis *et al.* (2015)
• Difficulty maintaining constant feed DSs to hydrolysis resulted in delay • Poor initial accuracy of online instruments required temporary alterations to control software • Pre-heated dilution water helps with energy balance	Jolly *et al.* (2012)
• Stratification of sludge in storage silos • Construction debris found inside pumps, import silo and other equipment causing internal damage • Quality and inconsistency of feed-sludge fundamentally influences start-up with slugs of biological or septic sludge causing digester upset during start-up • Sludge changing in quality due to seasonal variations	Belshaw *et al.* (2013)
• Incident with high pressure causing damage • Incorrect rupture disks/components	van Veldhoven and Smits (2019)

(Continued)

Table 4.1 Observations during start-up and commissioning of thermal hydrolysis plants with digestion (*Continued*).

Comments	Reference
• Sludge in condensate piping • Sludge overflows • Potential for high H_2S in biogas • Cooling capacity of heat exchanges inadequate • Clogging of heat exchangers • Temperature profile across equipment • Foreign objects and stones in sludge • Foul gas composition • Clogging of extruder • Maintenance of correct solids in pre-thickening centrifuges • Start-stop issues with pre-thickening centrifuges	
• Need to keep consistent feed • Be aware of fluctuations in primary:biological sludge ratio	Anglian Water (2008)
• Build-up of inorganic particles from imported sludge cakes over time • Effluent used for cooling hydrolyzed sludge causes fouling in subsequent heat exchangers due to the presence of suspended solids and organic matter. This reduces cooling efficiency resulting in warmer sludge being fed to digesters than anticipated • Coagulants used in dewatering consume alkalinity which is required by downstream deammonification process which incurs unexpected operating cost for additional alkalinity supplementation	Wang *et al.* (2018)
• Fouling with vivianite prior to pre-dewatering plant if large quantities of ferric are used upstream	Pathak *et al.* (2018)
• Vivianite noticed in digestion due to ferric addition	Driessen *et al.* (2018)

4.2 RAPID RISE AND SLUDGE VOLUME EXPANSION

There have been several reports in the literature pertaining to a phenomenon known as rapid rise whereby stoppages in digester mixing can cause an increase in sludge levels within the digester. Being a shear-thinning thixotropic fluid, the viscosity of sludge decreases with increasing shear (Chapman & Krugel, 2011) and shown in Figure 3.1, therefore if mixing within the digester is reduced, viscosity of the contents increases which entrains biogas bubbles causing an increase in level. Thames Water have reported on this phenomenon at several full-scale sites and noted that samples taken from digesters would decrease in volume by over 4% within only 5 min (Fountain & Mercedo, 2013). In the same work, it was stated that volume expansion due to thermal hydrolysis would result in digested sludge having a density between 0.85 and 0.93 t/m^3, compared with digested sludge with no pre-treatment which would remain similar to that of water. It is common to see an almost instantaneous increase in biogas production when mature thermally hydrolyzed sludge-fed digesters receive sludge. Therefore, intermittently fed digesters are at risk of sudden changes in sludge volume with potentially negative implications. In later research, a team led by Higgins *et al.* (2017) showed the influence of sludge volume expansion by switching off digesters which were fed thermally hydrolyzed sludge with hydrolysis reaction temperatures between 130°C and 170°C. The work showed decreasing volume expansion with increasing reaction temperature from 6% volume increase at 130°C, to <3% at 160°C. No expansion was noted at all at 170°C. At typical operating temperatures of thermal hydrolysis, the expected volume expansion would be approximately 2% (Higgins *et al.*, 2017). These results are consistent with higher viscosities and shear strength typical of lower reaction temperatures. As reaction temperature increases, viscosity decreases resulting in less entrapment of biogas and concomitantly less volume increase.

Measures which can be taken to decrease the impacts of volume expansion during digestion of thermally hydrolyzed material include:

- Design digester volume to account for lower sludge density
- Add additional headspace to collect expanded sludge if necessary; approximately 10–15% is sufficient
- Avoid, where possible, intermittent feeding of digesters
- Avoid, where possible, sudden changes in sludge composition
- Backup generators to avoid mixing stoppages may be prudent if none of the above measures are accounted for, otherwise they may be unnecessary

4.2.1 Composition of biogas and off-gas

The biogas composition during anaerobic digestion is controlled by stoichiometry (Buswell & Neave, 1930 and subsequent text; McCarty, 1971; Rittmann & McCarty, 2001), consequently, is independent of pre-treatment. However, minor differences are expected due to changes in the digester's chemistry. As digesters run at higher pH with thermal hydrolysis, due to concentrated biomass, enhanced deamination and concomitant generation of alkalinity, sulfide equilibrium progresses toward HS^- and S^{2-} from $H_2S_{(aq)}$. Subsequently, sulfide compounds in the biogas are lower in the presence of thermal hydrolysis. Wilson's team (2011) showed that total volatile organic sulfur compounds in the headspace were typically 10–15% of the results measured for control digesters with no pre-treatment. However, the off-gas produced by thermal hydrolysis has hydrogen sulfide within it (*see below*), and if not adequately managed can be measured in biogas and potentially be problematic. The increased pH and elevated TAN levels cause ammonia in solution to migrate toward $NH_{3(aq)}$ which is then in equilibrium with the headspace biogas. From typical performance and digester pH, theoretical ammonia concentrations in biogas should be in the range of 170–240 ppmv in the biogas. Measurements imply that a fraction of this remains absorbed such that 50–85 ppmv are observed in the biogas. This is three to four times higher than anaerobic digesters with no pre-treatment. In some instances, methane content appears higher, but this is expected due to changes in bicarbonate chemistry at a higher operating pH.

An off-gas is produced during the thermal hydrolysis process. Typically, this gas is collected, treated and fed to the anaerobic digestion plant. The gas consists of mainly carbon dioxide (85%), followed by nitrogen (~10%), oxygen (~1%) and the remaining 5% comprising hydrogen sulfide, carbon monoxide, hydrogen, trace quantities of nitrous and nitric oxide, and various volatile organic compounds. The following volatile compounds have been measured in the off-gas in concentrations greater than 1 ppmv: methylethylketone, 3-methylbutanal, acetone, 2-methylbutanal, isobutyraldehyde, methylmercaptan, dichloromethane and 3-methylthiophene.

4.2.2 Foaming

Foaming during wastewater treatment is undesirable and causes problems such as lost digestion capacity and increased maintenance and operational costs. The prerequisites for foam production of surface active agents, a gas phase and hydrophobic material are all present during anaerobic digestion and it is management of these parameters which will govern whether foaming will be a potential issue (Barber, 2005). The reported causes of foaming during anaerobic digestion include: presence of fats, oils, grease, soap, surfactants and waste-activated sludge in the feed stream; production of bio-surfactants; incorrect start-up of digestion; poor feeding control and/or intermittent feeding; use of biogas mixing; inadequate mixing; poor maintenance; excessive alkalinity and excessive use of polymer for pre-digestion thickening (Barber, 2005). In the review by Barber (2004), a mechanism for foaming during digestion is hypothesized based on the pertinent literature as follows:

- Disturbance in digester causes metabolic imbalance, which stimulates the excessive production of bio-surfactants and lysis products which drops surface tension causing a gradient and subsequent Gibbs–Marangoni effect
- A combination of bio-surfactants (and/or excessive volatile fatty acids caused by metabolic inhibition) and biogas result in the onset of foam
- Foam provides mechanism to cause proteins to unfold and/or denature, therefore enzymes using these proteins will lose their viability and cause further metabolic breakdown

- Under increasing disturbance, bacteria excrete extracellular polymers as a barrier to stress and to store electron donor for future use. These polymers act to reduce the liquid drainage from bubble lamellae by increasing viscosity, thereby stabilizing the foam

Thermal hydrolysis fundamentally influences the parameters which control foam formation and subsequent stabilization. Principally, a reduction in viscosity improves liquid drainage from bubble lamellae which reduces foam stability. By improving digestion performance, microbiological stress is reduced which corresponds to reduced production of extracellular polymers. Furthermore, thermal hydrolysis improves solubility of surface active compounds and hydrophobic materials which make the generation of a surface tension gradient more difficult. All of these parameters, in principle, make the onset, and subsequent stability of foam significantly less. However, increased gas mixing due to enhanced volatile solid removal, elevated alkalinity, production of melanoidins (Lusk *et al.*, 1995) and release of hydrophobic fragments are factors which all could enhance foam production and stabilization. Subsequently, there have been mixed opinions on the efficacy for thermal hydrolysis to help with foaming in the published literature. Typically, laboratory tests predict higher foaming potential, but this is inconsistent with full-scale data.

At lab-scale, digesters fed thermally hydrolyzed sludge were found to foam more than digesters with unpretreated material (Oosterhuis *et al.*, 2014) and the authors assumed this to be due to a combination of increased volatile fatty acids and biogas production. In a Spanish study, the influence of thermal hydrolysis (alongside thermal treatment <100°C, and ultrasound) on foam production and subsequent stability was studied (Alfaro *et al.*, 2014). The authors used a well-known test which is used for activated sludge systems. In the test, they added over-the-counter effervescent antacid tablets to the samples in a cylinder and then measured foam production over time. The volume of foam was a measurement of foam potential, while the length of time required half the foam to collapse was dubbed the foam stability. Data were collected prior to and post pre-treatment. Interestingly, the authors found that while foam potential decreased with increasing thermal hydrolysis temperature (between 120°C, 150°C and 170°C) it was no better than a control sample which had not been thermally adjusted. On the contrary, many of the test samples showed a higher propensity for foaming than the control. Timing of treatment between 15, 30 and 60 min was inconsequential. Furthermore, when foam stability was measured, similar trends were noted, whereby thermal hydrolysis resulted in foams lasting longer than the control. Data were lower in the absence of steam explosion (also tested as a variable) but remained worse. Results for ultrasound also followed similar patterns. The authors suggested that cell fragments were released from hydrophobic bacteria, and these retained their hydrophobic characteristics. Intrigued by the results—which were also inconsistent with their own data from various facilities, and studies by other authors (Barjenbruch & Kopplow, 2003; Marneri *et al.*, 2003)—the authors also looked at abundance of *Microthrix parvicella* (notorious for causing foaming events) which was identified in the activated sludge feed. They found that pre-treatment reduced abundance of these bacteria, and based on work by Westlund *et al.* (1996; cited by Alfaro *et al.*, 2014) discovered that several combinations of treatment would reduce the abundance of *M. parvicella* to below levels necessary to sustain foaming events. The authors concluded that thermal hydrolysis with steam explosion for (either 15 or) 30 min at 170°C, and ultrasound at a dose of 66.7 kWh/m^3 (using a piezoelectric device at 24 kHz) eliminated the potential for foaming from sludge. Temperatures and conditions which were less severe showed data which would support foaming (Alfaro *et al.*, 2014). The scientists suggested that while the methodology was acceptable for activated sludge monitoring, it was not viable for testing pre-treatment and digestion and that a new test was required. The testing results showed no correlation with the microbiological analysis.

The concerns of Alfaro's team over the ability of the testing protocol to provide accurate results were validated later by Higgins *et al.* (2017) who used the same methodology. The work of Higgin's team looked at sludge which had been both thermally hydrolyzed and digested, unlike Alfaro's coterie who only measured across the pre-treatment step. Although the later study had no control, nor looked at the impact of steam explosion or study foam stability, results for foam potential mirrored the previous work. However, in stark contrast (to Alfaro *et al.*, 2014), Higgins' work showed that foam potential become increasingly higher with increasing thermal hydrolysis reaction temperature. At full-scale, a reduction in thermal hydrolysis operating temperature—to potentially alleviate inhibition concerns in a downstream deammonification plant—from 165°C to 155°C resulted in the onset of foam in the anaerobic digester (de Clippeleir *et al.*, 2019). A gradual increase in temperature by 5°C showed decreased tendency for foaming. Unfortunately, it was not possible to sample the foam; however, it is likely that the foam was transient and associated with bacterial activity rather than metastable associated with bacterial stress based on the continued high performance and gas production measured within the digester. While the laboratory studies provide inconsistent results, data from full-scale facilities suggest that foaming potential during anaerobic digestion is significantly diminished when thermal hydrolysis is installed upfront of digestion; this is especially true if operating temperature is approximately 170°C.

4.3 RETURN LIQUORS FROM DEWATERING

4.3.1 Influence of thermal hydrolysis on nutrient solubilization

The combination of thermal hydrolysis and digestion yields liquors which are laden with nutrients, COD, solids and alkalinity. A fraction of the material is refractory as

> *Thermal hydrolysis can increase concentration of nutrients by over a factor of 3 in digestate*

discussed in Chapter 3, and can enter the final effluent of the treatment works. The impact of and cost of treating these liquors is an important consideration when weighing up the potential of thermal hydrolysis.

The work of Zhuo's coterie (2015) showed the influence of both thermal hydrolysis (165°C for 50 min) and digestion on the fate of nitrogen and phosphorous for both mesophilic and thermophilic systems. Figures 4.3 and 4.4 show the fate of nutrients for nitrogen and phosphorous, respectively (plot from data presented in Zhuo *et al.*).

As shown in Figures 4.3 and 4.4, the influence of solubilizing nitrogen appears more profound than phosphorous. The results showed that for sludge fed to digestion, the vast majority of the nutrients were not solubilized and contained within the sludge biomass itself, as one would expect. With nitrogen, solubilization was increased by over 10 times with the majority being measured as ammonia. This is consistent with the data presented by several workers who show that proteins are influenced by increasing temperature more than other compounds. By contrast, solubilization of phosphorous increased by a factor of 3–4 and comprised mainly of soluble phosphate.

Besides releasing more nutrients into solution, improved digestion performance downstream results in the additional release of nutrients. In a separate unpublished study (Li *et al.*), total nitrogen increased from 1,000 to 3,000 mg/l, and total phosphorous increased from 150 to over 250 mg/l for digested and thermally hydrolyzed digested sludge, respectively. Interestingly, while total phosphorous increased, the quantity which was orthophosphate was similar in both cases at circa 150 mg/l.

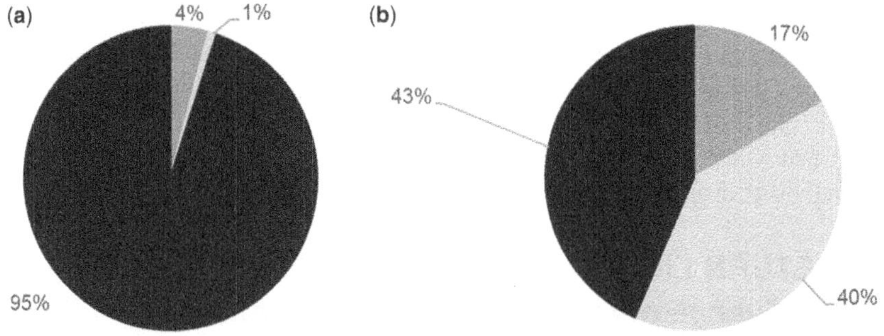

Figure 4.3 Influence of thermal hydrolysis (without digestion) on the solubilization of nitrogen: (a) before treatment and (b) after treatment. *Key*: Black = insoluble nitrogen; lightest gray = soluble organic nitrogen; gray = soluble ammonia nitrogen.

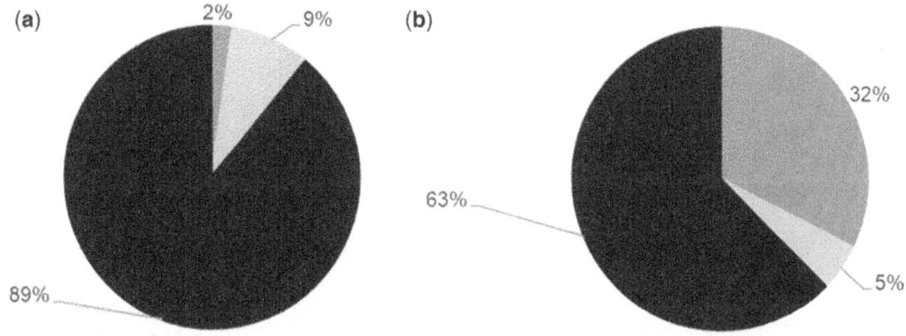

Figure 4.4 Influence of thermal hydrolysis (without digestion) on the solubilization of phosphorous: (a) before treatment and (b) after treatment. *Key*: Black = insoluble phosphorous; lightest gray = soluble organic phosphorous; gray = soluble phosphate.

Han's group (2017) measured the fate of proteins, nitrogen and phosphorous before and after thermal hydrolysis at 165°C for 50 min. Prior to thermal hydrolysis, sludge samples had 47 and 1.8 g/l particulate and dissolved protein, and these changed to a 1:1 ratio after heating. Looking closer at the nitrogen content, there was a large decrease in particulate organic nitrogen from almost 6 to 2.74 g/l. This reduction was combined with increases in dissolved organic nitrogen (DON) (0.06–2.51 g/l) and ammonia nitrogen (0.27–1.06 g/l). With respect to phosphorous, the influence of thermal hydrolysis was slight with respect to particulate levels with a decrease to 1.35 from 1.94 g/l (Han *et al.*, 2017). However, destruction of phosphorous-accumulating organisms observed a large increase in phosphate from 0.05 to 0.7 g/l. This was by far the highest influence of thermal hydrolysis on phosphorous balance, as dissolved organic phosphorous increased by a less amount (0.07–0.11 g/l; Han *et al.*, 2017).

4.3.2 Nitrogen

4.3.2.1 Ammonia

This compound gains much attention as it has a direct impact on the effluent discharge. Under anaerobic conditions, nitrogen contained within proteins is degraded to ammonia and alkalinity as shown in the following reactions (Speece, 2008):

$$RCHNH_2COOH + 2H_2O \rightarrow RCOOH + NH_3 + CO_2 + 2H_2 \quad (4.2)$$

$$NH_3 + H_2O \rightarrow NH_4^+ + HCO_3^- \quad (4.3)$$

As shown in equation (4.3), for every unit of mass of nitrogen released during digestion, 3.6 units of alkalinity expressed in calcium carbonate equivalents is released. Therefore, concentration of both nitrogen and alkalinity in the effluent

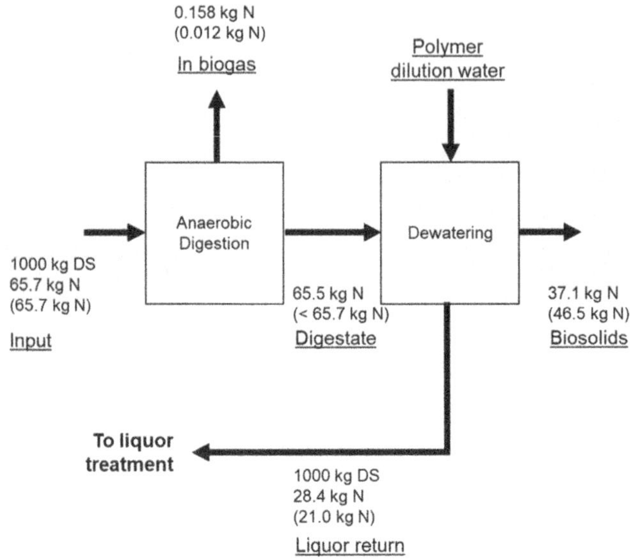

Figure 4.5 Nitrogen balance around anaerobic digestion and dewatering for 100 t/d DSs of sludge. *Key*: No parentheses = thermal hydrolysis precedes anaerobic digestion; in parentheses = no pre-treatment.

can be determined via mass balance with the knowledge of the nitrogen contribution of the volatile solid fractions and digester performance. Research has shown that approximately 15–20% of the ammonia released is absorbed and remains within the biosolids.

Figure 4.5 shows a mass balance for nitrogen around anaerobic digestion and dewatering for 100 t DSs/d where digestion is preceded by thermal hydrolysis or not (shown in parentheses in Figure 4.5).

The quantity of ammonia released is dependent, therefore, on the performance of anaerobic digestion, as demonstrated in the following example.

Question. How much ammonia is released from thermal hydrolysis and digestion of 100 t DS/d sludge containing 75% volatile material of which 7% is nitrogen. Assume volatile solids destruction is 60%?

Answer.
The quantity of nitrogen entering the digester

$$= 100 \text{ (t DSs/d)} \times 75\% \text{ (Volatitle fraction)} \times 7\% \text{ (nitrogen)} = 5.25 \text{ t}$$

If this is converted to ammonia and 20% remains within the biosolids, then the release of nitrogen is

$$= 5.25 \text{ (t/d)} \times 60\% \text{ (volatile solids destroyed)} \times (1 - 0.2) \text{ (released)} = 2.52 \text{ t}$$

If the digester is being fed at 10% DSs, this is equivalent to a concentration of 2,520 mg/l. Also, from equation (4.2), alkalinity released would be approximately 9,100 mg/l. These are the expected concentrations from digestion; however, they will be diluted depending on how much water is supplemented for polymer addition in the subsequent dewatering stage. Generally, these values will reduce by 25–30% after water addition.

With knowledge of digester temperature, equilibrium data and pH, it is possible to determine free ammonia concentrations and quantity of ammonia in the headspace of the biogas. The following equation can be used to determine free ammonia nitrogen (FAN) from TAN (Hansen *et al.*, 1998):

$$\text{FAN} = \text{TAN}\left[1 + \frac{10^{-\text{pH}}}{10^{-(0.09018+(2729.92/T))}}\right]^{-1} \tag{4.4}$$

where FAN is free ammonia nitrogen, TAN is total ammonia nitrogen and T is the temperature (°K). Figure 4.6 compares typical ammonia concentrations for standard anaerobic digestion with digestion preceded by thermal hydrolysis of all sludge or only the waste-activated sludge fraction.

The graph shows an increase in TAN concentration of between two and three times based on sludge composition. The increase reduces with increasing concentrations of primary sludge due to primary sludge contains fewer nutrients than activated sludge, and lower influence of thermal hydrolysis on improvement in digestion performance.

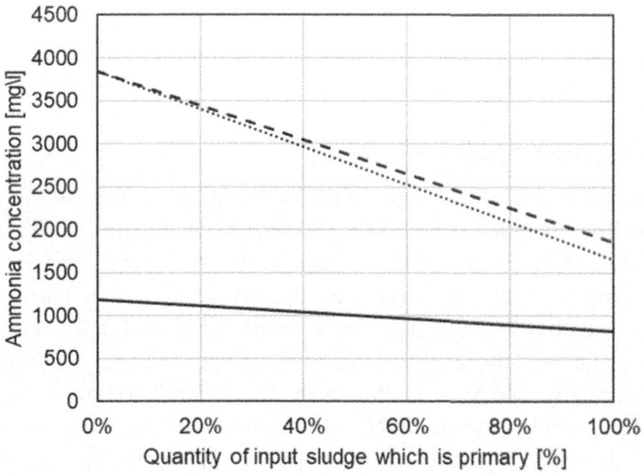

Figure 4.6 Influence of sludge composition on the predicted release of TAN from anaerobic digestion. *Key*: Full-line = no thermal hydrolysis; dashed-line = thermal hydrolysis of both primary and activated sludge; dotted-line = thermal hydrolysis of only activated sludge.

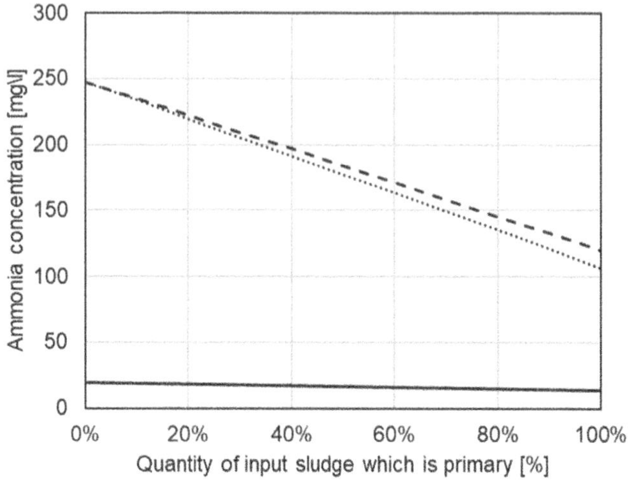

Figure 4.7 Influence of sludge composition on the predicted release of free ammonia from anaerobic digestion. *Key*: Full-line = no thermal hydrolysis; dashed-line = thermal hydrolysis of both primary and activated sludge; dotted-line = thermal hydrolysis of only activated sludge.

However, when free ammonia is determined from typical operating conditions within the digester, the influence of pre-treatment is far more prominent as shown in Figure 4.7.

With unassisted digestion, free ammonia lies between 20 and 30 mg/l. However, due to increased reactor pH and elevated digestion temperature, thermal hydrolysis digestion increases free ammonia between 7 and 13 times to between 100 and 250 mg/l depending on sludge composition.

4.3.2.2 Dissolved organic nitrogen

Various researchers have measured DON in the range of 400–800 mg/l (Dwyer *et al.*, 2008; Phothilangka *et al.*, 2008; Ahuja, 2015; Higgins *et al.*, 2017). In a Chinese study (Zhuo *et al.*, 2015), composition of nitrogen species following thermal hydrolysis with digestion was approximately 30% particulate, 70% dissolved of which 80% was ammonia (both TAN and FA). Therefore, the remainder was DON. This suggests that DON is approximately 25% of the TAN measured. Typically, TAN is in the range of 2,400–2,800 mg/l based on the Chinese study, would result in DON levels of 600–700 mg/l. In comparison, the equivalent values for digestion with no pre-treatment showed an equal share between particulate and dissolved nitrogen, of which 60% was TAN. Higgins' group (2017) also showed data where DON was approximately 30% of the nitrogen present. Li's team (2016) showed total nitrogen in effluent of approximately 3,000 mg/l of which ammonia-N was approximately 2,400 mg/l, making DON approximately 600 mg/l. The work by Ahuja (2015) shows that DON is dependent on thermal hydrolysis reaction temperature with results of approximately 340, 615 and 765 mg/l DON for reaction temperatures of 130°C, 150°C and 170°C, respectively. These findings were mirrored later (Higgins *et al.*, 2017) where a linear correlation was noted in DON production between a similar temperature values. In that work, DON in the digestate was approximately 500 mg/l independent of initial reaction temperature.

4.3.3 Phosphorous

Unlike nitrogen, the understanding of the fate of phosphorous has gained little interest in research areas with less published information. Similar to nitrogen, the study of Zhuo's team (2015) also monitored the fate of phosphorous across thermal hydrolysis with and without digestion. They found little difference due to thermal hydrolysis with respect to the particulate and dissolved phosphorous in digestates with approximately 55–60% being particulate and the remainder dissolved. However, in the digester with no pre-treatment the dissolved fraction contained approximately 75% phosphate, this increased to almost 95% when thermal hydrolysis was present. Molokwu and Rus (2017) showed a variation in soluble phosphorous in the digestate of full-scale plants with thermal hydrolysis. The studies of both Beckton and Crossness treatment had approximately 200 mg/l soluble phosphorous in the feedstock to digestion and approximately 80% of this value was in the effluent. However, at Oxford, Crawley, Longreach and Chertsey—where input phosphorous varied from 50 to 150 mg/l—less than 30% of the soluble phosphorous was measured post digestion. The authors did not delve into the speciation of the soluble phosphate. However, Li *et al.* (2016) showed total effluent phosphorous of circa 270 mg/l of which half was orthophosphate. Driessen's group (2018) found that soluble phosphorous concentration in filtrate increased from 10 to 75 mg/l after upgrading an existing anaerobic digestion plant to one with thermal hydrolysis pre-treatment.

Han's team (2017) studied the influence of phosphorous across various process steps inclusive of thermal hydrolysis and digestion. They concluded that thermal treatment had little influence on the transformation characteristics of phosphorous from subsequent anaerobic digestion—similar to findings of Zhuo *et al.* (2015). The results showed that between 32% and 35% of total phosphorous in the sludge was transformed to phosphate (Han *et al.*, 2017). To elucidate the source of phosphorus post digestion, phosphate release tests on the waste-activated sludge prior to processing were conducted. A phosphate/total phosphorous ratio of approximately 30% was noted and this was very similar to the ratio of the digested material (Han *et al.*, 2017). The authors concluded that phosphate produced during digestion was a consequence of polyphosphate hydrolysis. Due to increased sludge concentration and solubilization, phosphorous recovery appears to be a complementary with thermal hydrolysis and digestion, and has recently been gaining traction (Driessen *et al.*, 2018; Taylor, 2019).

4.3.4 COD

As with nitrogen, COD in the effluent can be determined by a simple mass balance based on the COD:VS ratio of the incoming sludge and the level of COD destruction to form biogas. This is shown by means of an example.

Question. Assume 100 *t DSs are digested with a volatile solid content of* 75% *and a COD:VS ratio of* 1.6. *Sludge is digested at* 10% *solids and* 62% *of the COD is converted to biogas. What is the total COD concentration exiting the digester? Approximately* 42,000 m^3/d *of biogas are produced. Determine the biogas production based on the destruction of COD.*

Answer.
The sludge contains 100 (t/d) × 75% (volatile fraction) = 75 t volatile material/d.

With a COD:VS ratio of 1.6, this is equivalent to 120 t COD/d.

It is fed at 10% DSs, therefore there are 120 t COD in 1,000 m^3 sludge/d. This is a concentration of 120,000 mg/l of total COD.

In the digester, 62% of the COD is destroyed, i.e., 74.4 t/d. This leaves an effluent containing 120 (t/d fed) – 74.4 (t/d destroyed) = 45.6 t/d. As before, with a flow of 1,000 m^3/d this equates to a concentration of 45,600 mg/l.

The biogas production can be simply determined from stoichiometry whereby 1 kg $COD_{destroyed}$ produces 0.35 m^3 methane at standard temperature and pressure.

Therefore, if 74.4 t COD are destroyed, methane production

$$= 74.4\ (t/d) \times 1{,}000\ (kg/t) \times 0.35\ (m^3/kg)$$
$$= 26{,}040\ m^3/\text{methane}.$$

Assuming a biogas methane content of 65%, this is equivalent to a biogas production of 40,062 m^3/d, which is similar to the measured production.

Subsequently, depending on the VS:COD ratio of the sludge, effluent total COD is in the range of approximately 40,000 to 60,000 mg/l. As for the nitrogen, this concentration is further diluted by 25–30% due to the addition of water for polymer makeup. Data collected from site and laboratory scale show that between 15% and 35% of the COD leaving the digester is soluble, and of that fraction approximately a third is biodegradable. Therefore, in this example, with a total COD of 45,600 mg/l (assuming 20% is soluble), approximately 9,120 mg/l is soluble and a little over 3,000 mg/l is biodegradable. A study investigating the fate of nutrients with and without thermal hydrolysis prior to digestion (Li *et al.*,

2016) showed a digestate COD level of over 7,000 mg/l. Approximately a third was soluble of which 40% was biodegradable. Denitrification requires between 3.5 and 4.0 mg/l biodegradable carbon/kg ammonia-N removed (Daigger, 2014), therefore in this example, there is sufficient carbon to denitrify 860 mg/l (approaching a third of the) ammonia-N. It is the presence of this material which interferes with deammonification by stimulating the growth of heterotrophic bacteria (Gu *et al.*, 2018) causing bacteriostatic inhibition (*see later*).

A Dutch study showed that the inert fraction of COD in the digestate was equivalent to approximately 11 kg COD/t DS (Oosterhuis *et al.*, 2014). Based on typical mixed sludge with a VS content of 75% and a COD equivalence of 1.5 and expected COD destruction within the digester, this is approximately 1% of the incoming total COD, or almost 2.5% of the COD exiting the digester.

4.4 TREATMENT OF RETURN LIQUORS

Although return flow from dewatering is low with respect to plant influent, the loads of nutrients returned can be significant. Figure 4.8 shows the influence of (a) flow and (b) load of nitrogen based on thermal hydrolysis and digestion.

As Figure 4.8 clearly shows, in terms of return flow, the quantity is negligible, in this case 0.5% of the influent. However, the return nitrogen is 21% of the total entering the works. For a plant processing only activated sludge this decreases slightly to 20% of the influent quantity. In comparison, the calculated values for standard digestion are a return flow of 0.9% containing 16% of influent nitrogen equivalents.

Typically, treatment of liquors can be via the use of spare capacity of existing biological treatment works, or involve a purpose-built high rate treatment plant—typically deammonification, or use other means, such as evaporation, and all of these options are operating with thermal hydrolysis and digestion at full-scale. Alternatively, existing infrastructure can be modified and repurposed (Hollowed *et al.*, 2019). A number of proprietary high-rate biological systems have been used with thermal hydrolysis including: the following deammonification systems, Anammox® (Paques), Anita™Mox (Veolia), Demon® (World Water Works), Cleargreen™ (Suez), Renocar (developed by Beijing Drainage Group), and the following different high-rate systems, Amtreat® denitrification with nitrification (ACWA) and SHARON (Single-reactor, High-Activity Ammonia Removal Over Nitrite).

In principle, the digestate produced from digesters fed thermally hydrolyzed sludge is not compatible with the use of deammonification. As with potential inhibition from refractory compounds (Section 3.3) the digestate has a high COD/N ratio and elevated suspended solids which encourages the growth of heterotrophs that outcompete the chemoautotrophic bacteria (Gu *et al.*, 2018), subsequently causing eventual washout due to their sluggish doubling times of 7–22 d (Kartal *et al.*, 2013). Furthermore, high COD/N ratio may encourage the

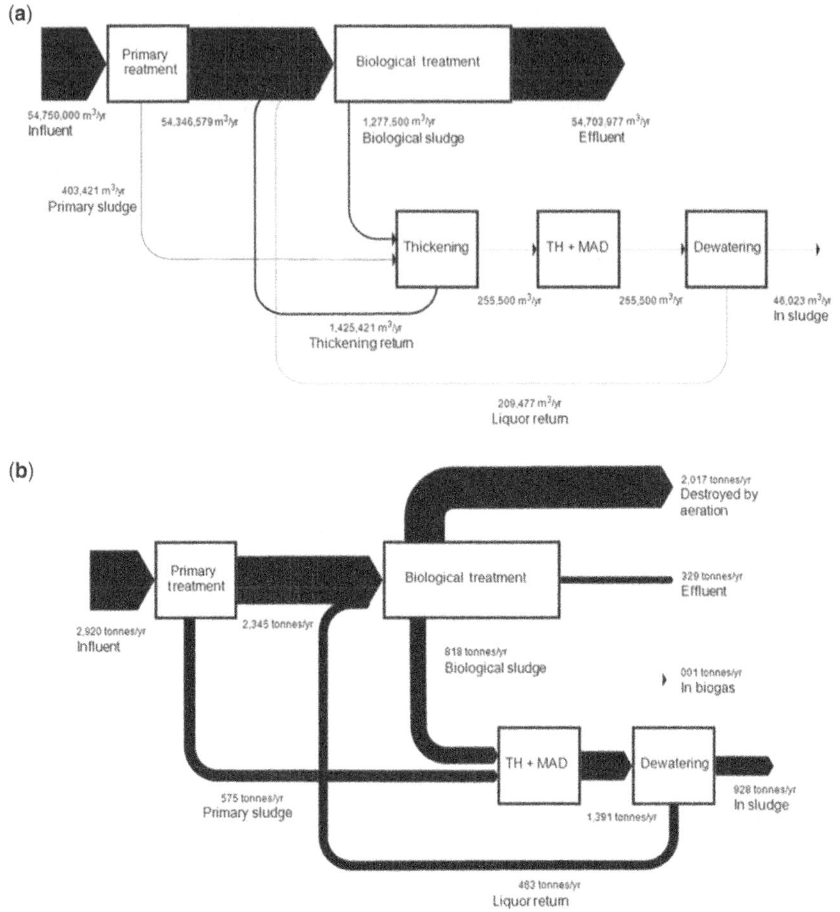

Figure 4.8 Influence of thermal hydrolysis and digestion on return flows and loads of nitrogen. Based on 1 million population equivalent (PE) plant, water consumption of 150 l/PE/d, 8 g nitrogen/PE/d. Based on thermal hydrolysis of both primary and activated sludge upstream of digestion.

alternative nitrogen pathway of dissimilatory nitrate reduction to ammonium (DNRA). The trophic groups catalyzing DNRA and anammox pathways have been found to coexist, however, DNRA bacteria can out-compete anammox by altering the optimum nitrite/ammonia ratio causing washout of anaerobic ammonia-oxidizing bacteria (AnAOB; Gu *et al.*, 2018). However, when total nitrogen removal is required, these systems become attractive due to lower operating costs than traditional activate sludge treatment, due largely to the savings from carbon (e.g., methanol) supplementation. Deammonification systems need the digestate to be diluted with water by a ratio of 1–1.5 units of water per unit of digestate treated to overcome heterotrophic interference (Driessen *et al.*, 2018; de Clippeleir *et al.*, 2019). Alternatively, deammonification systems may have pre-treatment to remove COD and suspended solids.

Driessen's team in the Netherlands (2018) presented 1 year of operating data and experience on the performance of an Anammox™ plant treating thermal hydrolysis digestion liquors. The plant also has Phospaq™ for phosphorous management. The Anammox™ facility was designed to remove a daily load of 1,850 kg. As anticipated, following an upgrade to advanced digestion, concentrations of COD, NH^4-N and PO_4 increased from 850, 500 and 10 mg/l up to 5,000, 1,850 and 75 mg/l, respectively. Although ammonia concentration fluctuated between 800 and 1,800 mg/l, performance was good with effluent remaining below 200 mg/l with removal rates in excess of 85%. New growth of anammox bacteria were noted after 150 d with granules of 2–4 mm in diameter (Driessen *et al.*, 2018). However, several, largely manageable operating issues were encountered. Overdosing polymer and poor control of solid capture resulted in foaming and deterioration of deammonification performance. The alkalinity/ammonia-N ratio declined over time which impacted the capacity of the system. Furthermore, there were concerns over *Legionella* (van Veldhoven & Smits, 2019).

Hollowed and team (2019) highlighted the performance of an Anita™Mox plant at Sundet WWTP (Växjö, Sweden) to treat digestate following an upgrade to thermal hydrolysis in 2014/5. The facility was converted from a moving bed biofilm reactor (MBBR) into an integrated fixed-film-activated sludge (IFAS) configuration to more effectively treat the higher strength filtrate. The facility also has co-digestion. The Anita™Mox system was retrofitted within an existing sequencing batch reactor (SBR) reactor and filled with 150 m^3 of AnoxK™ K5 media for a design load of 320 kg NH_4-N/d. After a period of inconsistent loading rates and variable performance, as with the work of Driessen *et al.* (2018), the plant was capable of removing 85% of the ammonia and over 75% of the total nitrogen similar to an MBBR system. After further upgrade to an IFAS configuration, performance improved further to <90% and <85% for ammonia and total nitrogen, respectively. The surface area loading rate increased to an average of 1.4 $g/m^2/d$ with the removal rates of 1.2 and 1.16 $g/m^2/d$ for ammonia-N and total nitrogen, respectively (Hollowed *et al.*, 2019). As with experience elsewhere, it was necessary to add dilution water. The authors speculated that expected microbial populations were suboptimal due to the ratio of nitrite/ammonia-N which deviated from the expected.

Blue Plains, in Washington DC has a DEMON™ plant comprising 5 × 3,000 m^3 reactors which has been operating since July 2018, with a capacity to treat approximately 5,000 kg/d (de Clippeleir *et al.*, 2019). This loading rate was achieved in three and a half months. As with other facilities, the filtrate is diluted using plant effluent at a dilution rate of approximately 1:1. Initial piloting experienced a reduction in performance, and the plant owners lowered the thermal hydrolysis operating temperature to alleviate perceived inhibition from refractory compounds as biodegradability of the soluble COD in the reactors fell from approximately 45% to 20%. However, a reduction in reactor temperature coincided with foaming (see earlier); therefore, reaction temperature was increased once more in the absence of any impacts on digestion, dewatering or deammonification. Performance at the plant has been good, with approximately 85% of the ammonia-N being removed. In a separate study based on the same plant, Zhang *et al.* (2018) looked closer at the mechanisms of the observed inhibition, and they also tried several methods of abatement. The authors found that the filtrate caused direct inhibition from biodegradable dissolved organic compounds on aerobic ammonia-oxidizing bacteria (AerAOB); however, these organisms were less sensitive to that inhibition than their anaerobic counterparts. The authors suggested that the inhibition could be mitigated by extended biological treatment (although this would fundamentally change alkalinity, pH, COD and ammonia levels) or use of activated carbon. By studying the impact of various COD fractions, Zhang's team found that particulates and large colloidal material inhibited deammonification via diffusion, and that activity of key bacteria was intrinsically linked to dewatering performance. An optimal dose of the polymer FLOPAM (10 kg/t DS) increased activity of AerAOB to almost 100%. Furthermore, when combined with 2 kg/t DS, polyDADMAC could achieve similar results irrespective of digester performance. By contrast, the optimum FLOPAM dose could only achieve a 20% activity for AnAOB, compared with 10% and 40% activities for under (7 kg/t DS) and over (14 kg/t DS) doses, respectively (Zhang *et al.*, 2018). During stable digestion performance, there was little significance in data for the use of FLOPAM (exclusively, combined with 2 kg/t DS polyDADMAC, or 0.07 kg/t DS ferric chloride) on activity of AnAOB which remained approximately 50%.

One study looked specifically at the influence of thermal hydrolysis on metabolic pathways occurring during the anammox process (Gu *et al.*, 2018). The authors found that COD levels were fundamentally important due to the response of heterotrophic growth at the expense of anammox activity. The work defined a specific anammox activity as the ratio of maximum substrate activity to VSS content, and this was used to determine the activity of AnAOB based on COD levels present. The work found no influence of the addition of 50 mg/l COD, but found a reduction in the activity of approximately a quarter at 100 mg/l and three quarters at 200 mg/l. The response is fitted to Luong inhibition model as shown in the below equation (taken from Gu *et al.*, 2018):

$$\text{Anammox activity} = \frac{2.95156 \times 10^6}{I^{3.06099} + 2.95156 \times 10^6} \tag{4.5}$$

where Anammox activity is expressed as a decimal and I is the concentration of COD in mg/l. In a subsequent part of the study, Gu's team looked at long-term anammox treatment using a membrane bioreactor. However, the authors found that it was not possible for the anammox organisms to acclimatize to the higher COD levels, indicating that the inhibition was bacteriostatic and would only be alleviated by reduction of the COD, as other teams have experienced. Inhibition in the reactor treating hydrolyzed material meant that the maximum volumetric loading rate was routinely 3.4 kg/m^3 d lower than untreated sludge; furthermore, as noted by other workers at full-scale (Hollowed *et al.*, 2019), the ratio of nitrite/ammonia drifted far from the ideal levels. The study identified that the predominant anammox contributor was *Candidatus brocadia*. Interestingly, it featured more in samples which had not been thermally hydrolyzed with levels between 4.35% to 8.3% and 1.94% to 2.36% of total counts in samples without and with thermal hydrolysis pre-treatment. The authors noted that previous literature showed that *C. brocadia* was more competitive than other AnAOB in the presence of acetate. Evidence of DNRA pathways was evident from microbial analysis. These findings suggest that the use of thermal hydrolysis could result in a shift in populations of anammox bacteria to those which are less robust.

Recently, the owning and operating experiences of one of Beijing's five proprietary made RENOCAR deammonification facilities was presented (Wang *et al.*, 2018). As with previous experience, it was necessary to dilute the filtrate, in this case to keep ammonia levels between 1,000 and 1,200 mg/l. This was done using a combination of final effluent and warm water from heat exchangers in order to maintain optimum temperature range for the bacteria involved. Even after 15 months of operation, the system has not been able to treat the full amount of filtrate, also typical of other systems with no inoculum to aid start-up. The process is operated with an internal recycle ratio of 3:1 and a sludge recycle of 1:1 as an IFAS system (similar to Växjö, Sweden). Average nitrogen removals for 10 months of operation were recorded as 92%, 88% and 87% for ammonia-N, total inorganic nitrogen and total nitrogen, respectively. These corresponded to volumetric nitrogen removal rates of 0.41, 0.39 and 0.42 kg N/m^3 d, respectively (Wang *et al.*, 2018). The authors hypothesized that the improved performance of the facility compared with other deammonification processes was due to partial heterotrophically catalysed denitrification in the anoxic zone. Subsequently, volumetric loading rates are lower than other systems and work is ongoing to further optimize the process.

Lately (Inkpin *et al.*, 2019), data were shared on the start-up and commissioning of a DEMON® facility at Basingstoke's 53 t DS/d thermal hydrolysis plant. The deammonification unit took over 5 months to reach design load. It is currently working well with over 75% ammonia removal despite fluctuating loadings. However, Inkpin's team identified various lessons learnt during the commissioning period. Instrumentation was incorrectly installed, the angle of the sensors gave false readings which impacted performance. Foaming occurred several times due to high sensitivity to polymer overdosing, and this required the use of anti-foaming agents. However, these reduced oxygen transfer which alerted the blowers to ramp up thereby providing more agitation which further exacerbated foam production. Although, a feed tank was installed, it was deemed to be too small, therefore there was little buffer from fluctuations in the dewatering system which would quickly impact the performance of the deammonification plant. Finally, it was found that cyclones (used to retain anammox bacteria) were incorrectly installed causing further process issues; however, this was readily rectified (Inkpin *et al.*, 2019).

Deammonification has been shown to be a good way of treating liquors from thermal hydrolysis plants with ammonia removal rates of approximately 85%, as the process can reduce operating costs on total nitrogen removal over use of existing infrastructure. However, as noted above, various aspects need to be considered as follows:

- A fraction of the COD in the material is biodegradable and can remove a portion of the nitrogen via denitrification. This reduces the operating cost benefits of deammonification
- COD/N ratio in the feed is too high. Biodegradable COD should be below 500 mg/l (Gu *et al.*, 2018) to prevent direct inhibition of anammox pathways. Typically, it is necessary to dilute the feed material to deammonification by a ratio of 1–1.5 water to one input feed. Alternatively, a pre-treatment solids and COD removal stage could be installed, or ferric and similar compounds and coagulants used to precipitate out soluble COD
- Use of certain coagulants to manipulate the sidestream prior to treatment may consume alkalinity required by deammonification, which results in additional operating costs
- It is prudent to install sufficient digestate storage prior to deammonification to maintain constant flow to the nutrient removal plant and protect it from variability in dewatering performance
- Performance of upstream dewatering with special attention to polymer dose and solids capture is critical with respect to inhibition of deammonification. Inadequate control of these parameters can result in deterioration of nitrogen removal and foaming events
- There may be additional operating costs for antifoaming agents and de-scalers, although these may be minimized by good operating practice around the dewatering device
- Melanoidins have a high affinity for chelating nutrients which may influence the efficacy of supplement addition to enhance or maintain performance
- Alkalinity/ammonia-N ratio may reduce over time, limiting the capacity of the plant for nitrogen removal. Addition of alkalinity may enhance performance
- The risks of *Legionnaires* need to be considered
- The influence of nitrous oxide emissions is significant when determining the environmental impact of the process (Schneider *et al.*, 2011; Schoepp *et al.*, 2018)

Other than deammonification, which requires dilution of the digestate, it may be possible to use the COD within the digestate as a carbon source for denitrification (Bungay, 2018). At United Utilities' Leigh thermal hydrolysis plant, an Amtreat® plant is installed. It is a purpose-built high-rate plant which is based on standard nitrification with denitrification. Compared with deammonification systems, its loading rate is approximately double at 1.2 kg N/m^3 d compared with 0.5–0.6 kg N/m^3 d for anammox-based systems. Ironically, on some sites Amtreat® is installed as a precursor facility to deammonification. Designs are based on the addition of glycerol as a carbon source and sodium hydroxide for alkalinity supplementation. At Leigh, the flowrate is 800 m^3/d with approximately 800 mg/l ammonia, naturally diluted down by the preceding process steps, although this is not necessary. As is typical, influent ammonia fluctuates significantly, however effluent levels are approximately 2 mg/l. Suspended solids is reduced from 800 to $<$20 mg/l. A high removal rate for phosphorous was also noted at Leigh, although it was not expected (Bungay, 2018). Although the plant was designed for chemical addition, to date there has been sufficient carbon and alkalinity in the digestate to render its addition unnecessary.

In contrast to biologically based systems, the biowaste plant in Oslo is using a different approach. The facility treats 50,000 wet t/year of source separated household waste combined with a portion of biowaste. Biogas produced from the plant (equivalent to 40–50 million kWh; Román *et al.*, 2012) is upgraded to biomethane and ultimately used to power some of the city's buses. The centrate from the plant is processed by evaporation. This produces a high nitrogen product which is used as a fertilizer. Unfortunately, there is little published information on the performance. This plant is referenced further in Section 4.5. Another thermal hydrolysis and digestion plant in Oslo (VEAS) is using air stripping to produce various ammonium-based liquid fertilizers.

4.5 CO-DIGESTION

The addition of organic materials to sludge to generate renewable energy has attracted much attention at both technical and governmental levels. Wastes which would likely have been landfilled only to generate greenhouse gases can be digested to produce renewable energy and a digestate which can be used as a fertilizer. However, these benefits are accompanied by manageable levels of risk as follows:

- Technical: increased wear and tear, liquor treatment; biogas processing; dewatering; maintenance; ability and cost of grid connection
- Financial: gate fee, operating costs, type of contract, cost and time period of available finance
- Market: availability of feedstocks; competition; alternative outlets

- Societal: nuisance, perception, increased traffic movements
- Turbulence: fluctuation in utilities costs, gate fee erosion, loss or degradation of incentives, regulation changes

These risks must be accommodated to ensure a successful co-digestion project. In principle, thermal hydrolysis can be used to create spare capacity in the existing digestion plants which could be used for the addition of organic substrates for co-digestion. These materials can be added in several locations in the plant as shown in Figure 4.9.

Besides the locations shown in Figure 4.9, certain low-solid feedstocks can be used for thinning the hydrolyzed sludge prior to digestion. The benefits and challenges of these addition points are described in Table 4.2.

As with co-digestion generally, viability is not straightforward and is critically dependent on various parameters, most important of which is the feedstock itself. Numerous studies have shown the influence of thermal hydrolysis on the solubilization of proteins, carbohydrates and lipids (Li and Noike, 1992; Bougrier *et al.*, 2008; Lu *et al.*, 2014; Xue *et al.*, 2015), and solubilization of these materials plays an important role when considering thermal hydrolysis of material other than sludge. Ironically, substances high in protein, such as activated sludge, are better suited to thermal hydrolysis than carbohydrates and lipids (see Chapter 1), but these typically yield less biogas and will release more nutrients which will incur a processing cost. Substances which are generally considered good

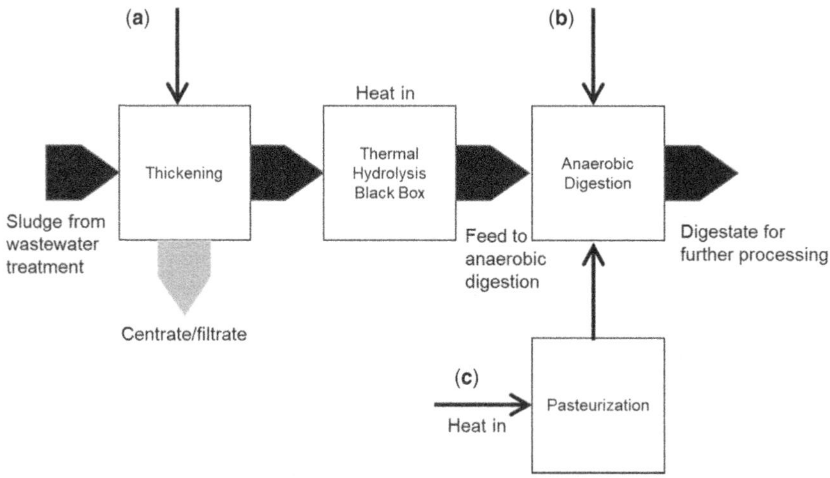

Figure 4.9 Potential addition points for organic waste in combination with thermal hydrolysis pre-treatment. (a) In with sludge prior to thickening and thermal hydrolysis; (b) directly into anaerobic digester and (c) into digester via the pasteurization process.

Table 4.2 Addition points for co-substrates for thermal hydrolysis and co-digestion.

Addition Point	Benefits	Challenges
Prior to thermal hydrolysis	• All material is sterilized • Biodegradability of certain substrates is improved	• Loss of solubilized material into centrate which then becomes a cost to treat when recycled to biological treatment • Material recycled in centrate material may cause operational issues (such as foaming) during biological treatment • Larger thermal hydrolysis plant • May change energy balance
As material to dilute sludge prior to digestion	• Saves on dilution water • No influence on size of thermal hydrolysis plant	• Needs to be sterilized, or enhanced treatment standard is lost • Temperature needs to be stable or will adversely impact cooling heat exchangers
Directly into digester	• All COD is captured in digestion • No influence on sizing or energy balance of thermal hydrolysis • Simple retrofit	• Enhanced treatment lost • May lose biogas yield from certain substances
Into digester via pasteurization	• Benefits as for adding directly into digester • Maintains enhanced treated standard	• Needs more equipment • Additional operating costs for pasteurization plant

candidates for co-digestion, i.e., those high in carbohydrates and fats, are less amenable to the influence of thermal hydrolysis.

As previously mentioned in Chapter 3, increasing protein and carbohydrate concentration at elevated temperatures encourages production of refractory material in the form of Maillard reaction products and caramelization products. In a detailed study, Tampio *et al.* (2014) discovered that biogas yield from food-waste co-digested with sludge was routinely 5–10% lower on a unit basis when compared with food-waste which was not thermally hydrolyzed under identical conditions independent of loading rate between 2 and 6 kg VS/m^3 d. The authors concluded that biogas production alone was an insufficient driver for the installation of thermal hydrolysis. This result complemented the findings of Liu *et al.* (2012), who investigated a wide variety of thermal hydrolysis settings with and without co-digestion of kitchen waste and vegetable fruit scraps with activated sludge. At 175°C/60 min—their default set-up—biogas yield reduced by approximately 15% and 10% for kitchen wastes and scraps, respectively. Data provided on UV absorbance at 254 confirmed a large increase in melanoidin across a molecular weight range between <1 and >300 kDa (Liu *et al.*, 2012). In a later work, Svennevik *et al.* (2019a, b) showed that foodwaste contained more refractory COD, dissolved organic carbon, higher UVA_{254} and color than sludge mixtures where over half the sludge was primary even though it was processed at only 145°C. But, it had lower numbers than high activated sludge content sludges processed at 165°C. In contrast, other work showed no significant reduction in biogas yield with pre-processed food-waste added at a rate of 25% by COD load (Barber, 2015) and that it was possible to predict biogas production based on stoichiometry and COD supplemented and destroyed. Depending on where the substrate is added and what it is, thermal hydrolysis with co-digestion may have the following influences:

- increased gas production but potentially with lower unit yield per quantity added
- increased steam demand
- treatment of hazardous materials in excess of regulatory requirements
- improved biogas production from certain difficulties to degrade materials
- enhanced digester stability (if there is a build-up of long and short chain fatty acids which depress pH and therefore reduce toxicity due to free ammonia)
- increased release of nutrients
- increased biosolids production
- increased polymer dose for dewatering
- similar dewatering

The operating conditions of thermal hydrolysis exceed the regulatory requirements of Europe's Animal Byproducts Regulations (Regulation No. 1069/2009) making the technology attractive for the processing of these types of materials. Despite the challenges of combining thermal hydrolysis with co-digestion, it has been successfully managed at full-scale at approximately 10 sites at the time of writing. Most existing plants and experience are in Northern Europe (Norway, Denmark and Sweden); however, there is increasing interest in South East Asia (Singapore and South Korea). Sludge to organic ratio has been from 0% to 85% addition with typical performance between 65% and 80% volatile solids destruction. Gas yields depend on substrate and vary from 540 to 800 N $m^3/t\ VS_{fed}$.

One interesting facility is in Oslo, Norway. This project was started up in 2013 to process food-waste at 28% DSs. Following anaerobic digestion, the waste is converted to liquid (using evaporation) and solid fertilizers, renewable energy and liquified biomethane which is used to run some of the City's buses. A brief overview is shown in Figure 4.10.

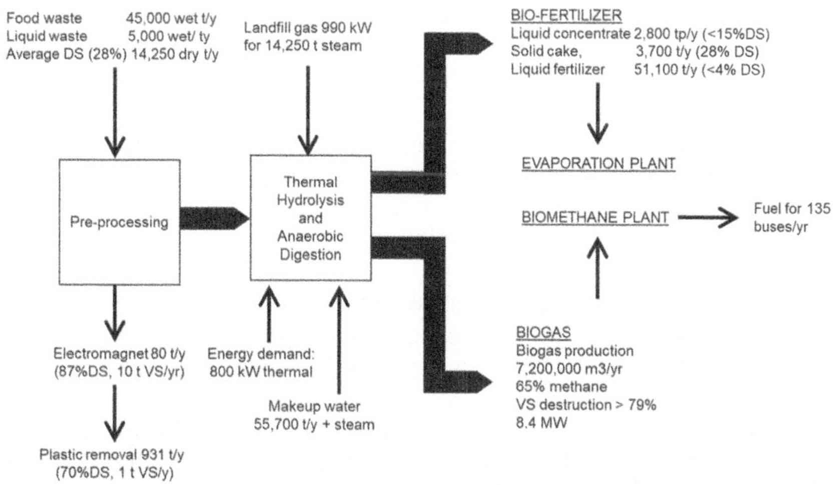

Figure 4.10 High-level mass balance for Oslo, source separated organics processing plant combining thermal hydrolysis with anaerobic digestion.

This facility is a good case-study showing a circular economy. Another excellent example, Billund in Denmark, is showcased in Chapter 6.

With full-scale operation, pre-processing the material prior to thermal hydrolysis is key to successful operation (Kanders & Sargalski, 2014). As with all co-digestion projects, when designing an appropriate pre-treatment system it is sensible to assume variability in the feedstock even if it is not anticipated.

4.6 POLYMER CONSUMPTION

The consumption of polymer, especially for thickening prior to thermal hydrolysis, is important when considering thermal hydrolysis. Polymer for thickening purposes can detrimentally influence the economics of thermal hydrolysis at a fundamental level (as shown in Chapter 7). Subsequently, the benefits of reduced energy requirements with a higher DSs feed to hydrolysis must be balanced against the increased costs of higher polymer dose. With respect to polymer dose post-digestion, thermal hydrolysis influences a number of parameters which have a direct influence. Work by Higgins and Novak (1997) presented a linear correlation between increasing protein concentration and increasing optimal polymer demand during dewatering. Other studies have shown polymer dose to be related to a higher concentration of fines with low particle size distribution, both of which are consequences of thermal hydrolysis. However, there are various counter arguments. Correlations have been shown which relate polymer consumption to volatile solids. Therefore, anything which reduces volatile content should decrease polymer dose. Molokwu and Rus (2017) studied the influence of extracellular polymer (ECP) on a variety of parameters known to influence both dewaterability and polymer consumption. They found a positive correlation between increasing ECP in digestate with increasing soluble phosphorous, mono to divalent cation ratio (M/D) and polymer consumption based on full-scale data from several thermal hydrolysis plants. They showed the following correlation with ECP:

$$\text{Poly dose kg/t Ds} = 2.6 \times 10^{-3} - \text{EPS (mg/1)} + 5.5 \tag{4.6}$$

Data from other full-scale plants show that polymer dose on digestion plants preceded by thermal hydrolysis is similar on a unit basis to plants with no pre-treatment. Lancaster (2015) showed a unit polymer dose of 8 kg active substance (AS)/t DS. The thickening dose was 6 kg/t DS. At DC Water (de Clippeleir, 2016) polymer dose was approximately 9 kg AS/t DS. However, van Veldhoven and Smits (2019) showed data from 16 to 18 kg AS/t DS for Zetag and 6.5 to 11 kg AS/t DS for Praestol. In a study comparing dewatering devices (Higgins *et al.*, 2011), polymer doses of 2.1–2.8 kg AS/t DS for a belt filter press and between 6 and 8 kg AS/t DS for a centrifuge were reported. However, polymer dosing can be higher during commissioning. It appears from full-scale data that polymer consumption is linked to sludge composition irrespective of the presence or absence of thermal hydrolysis, and that polymer demand increases with increasing proportion of waste-activated sludge.

In the absence of additional information, it is proposed that the polymer dose on a unit base of DSs is similar in the absence or presence of thermal hydrolysis, and highly dependent on site-specific conditions. Some facilities intentionally reduce polymer consumption to lower overall operating costs in-spite of a potential decrease in DSs in the final cake. Chapter 7 shows an example of the economics of doing this. A similar trend has been found when thermal hydrolysis is operated downstream of digestion. In a recent work (Svennevik *et al.*, 2019a, b), a linear trend was found between polymer dose and dewatering performance between polymer doses of 12 and 24 kg AS/t DS dewatered corresponding to dewatering potential between approximately 40% and 48%. The linear and high cationic charged polymer Zetag 9118 at a concentration of 0.2% active substance was used.

REFERENCES

Ahuja N. (2015). Impact of Operating Conditions on Thermal Hydrolysis Pre-treated Digestion Return Liquor, Thesis submitted to faculty of Virginia Polytechnic Institute and State University in partial fulfillment of the requirement for the degree of Masters of Science in Civil and Environmental Engineering.

Alfaro N., Cano R. and Fdz-Polanco F. (2014). Effect of thermal hydrolysis and ultrasounds pretreatments on foaming in anaerobic digesters. *Bioresource Technology*, **170**, 477–482.

Anglian Water. (2008). Anglian Water Biosolids Programme. Presented at Sludge Strategy and Advanced Digestion, Roundtable Event hosted by MWH at United Utilities.

Barber W. P. (2004). Foaming during anaerobic digestion. Causes and possible solutions, Paper, Proceedings of the Aqua Enviro/CIWEM 9th Biosolids and Organic Residuals Conference, Wakefield, UK.

Barber W. P. (2005). Anaerobic digester foaming: causes and solutions. *Water*, **21**, 45–49.

Barjenbruch M. and Kopplow O. (2003). Enzymatic, mechanical and thermal pretreatment of surplus sludge. *Advances in Environmental Research*, **7**(3), 715–720.

Belshaw D., Edgington R. M. and Jolly M. (2013). Commissioning of United Utilities' thermal hydrolysis digestion plant at Davyhulme Waste Water Treatment Works. Proceedings: 18th European Biosolids and Organic Resources Conference, Aqua Enviro Technology Transfer, UK.

Bougrier C., Delgenès J. P. and Carrère H. (2008). Effects of thermal treatments on five different waste activated sludge samples solubilisation, physical properties and anaerobic digestion. *Chemical Engineering Journal*, **139**(2), 236–244.

Bungay S. (2018). Treatment of Thermal Hydrolysis Liquors using the AMTREAT® Process. Proceedings of European Biosolids and Organic Resources Conference, November 2018, Manchester, UK.

Buswell A. M. and Neave S. L. (1930). Laboratory studies of sludge digestion. Bulletin (Illinois State Water Survey) no. 30.

Chapman T. and Krugel S. (2011). Rapid Volume Expansion – An Investigation into Digester Overflows and Safety. Proc. WEF Residuals and Biosolids Conference.

Daigger G. T. (2014). Oxygen and carbon requirements for biological nitrogen removal processes accomplishing nitrification, nitritation, and anammox. *Water Environment Research*, **86**(3), 204–209.

De Clippeleir H., Miranda M., Zhang Q., Hasan M., Al-Omari A., Passarelli N. and Debarbadillo C. (2016, January). Optimization of Belt Filter Press Dewatering of Thermally Hydrolyzed Biosolids: A Blue Plains Case Study on Balancing Desired Cake Solids with Filtrate Quality. Proceedings of the 89th Annual Water Environment Federation Technical Exhibition and Conference, pp. 26–30.

de Clippeleir H., Zhang Q., Wett B., Martinelli S., Miranda M., Cusic C., Kharhar S. M., Suzuki R., Passarelli N., Al-Omari A. and deBarbadillo C. (2019). Balancing physics, chemistry and microbiology in intensified biosolids and nutrient treatment systems. Proceedings of Water Environment Federation WEFTEC Conference.

Driessen W., van Veldhoven J. T. A., Hendrickx T. and Van Loosdrecht M. C. M. (2018). Successful treatment of sidestream dewatering liquors from thermally hydrolysed and mesophilic anaerobically digested (THP-MAD) biosolids. Proceedings of IWA Nutrient Removal and Recovery Conference, 18–21 November, Brisbane, Australia.

Dwyer J., Starrenburg D., Tait S., Barr K., Batstone D. J. and Lant P. (2008). Decreasing activated sludge thermal hydrolysis temperature reduces product colour, without decreasing degradability. *Water Research*, **42**(18), 4699–4709.

Fountain P. and Mercedo F. (2013) Practical Considerations for Feeding of THP Digestion. Proceedings of the 18th European Biosolids and Organic Resources Conference, UK.

Fountain P., Macedo F. and Shana A. (2015). A practical design guide for commissioning of THP digestion. Proceedings of 20th European Biosolids and Organic Resources Conference & Exhibition.

Gerardi M. H. (2003). The Microbiology of Anaerobic Digesters. John Wiley & Sons, Inc., Hoboken, New Jersey.

Gu Z., Li Y., Yang Y., Xia S., Hermanowicz S. W. and Alvarez-Cohen L. (2018). Inhibition of anammox by sludge thermal hydrolysis and metagenomic insights. *Bioresource Technology*, **270**, 46–54.

Han Y., Zhuo Y., Peng D., Yao Q., Li H. and Qu Q. (2017). Influence of thermal hydrolysis pretreatment on organic transformation characteristics of high solid anaerobic digestion. *Bioresource Technology*, **244**, 836–843.

Hansen K. H., Angelidaki I. and Ahring B. K. (1998). Anaerobic digestion of swine manure: inhibition by ammonia. *Water Research*, **32**(1), 5–12.

Hickey R. F., Wu W.-M., Viega M. C. and Jones R. (1991). Start-up, operation, monitoring and control of high-rate anaerobic treatment systems. *Water Science and Technology*, **24**, 207–255.

Higgins M. J. and Novak J. T. (1997). Characterization of exocellular protein and its role in bioflocculation. *Journal of Environmental Engineering*, **123**(5), 479–485.

Higgins M., Murthy S. N., Schafer P., Cooper A., Kasirga E., Lee K. and Kelleher K. (2011). Dewatering characteristics of cambi thermal hydrolysis biosolids: centrifuges vs. BFPs. Proceedings of the Water Environment Federation, **2011**(17), 678–684.

Higgins M. J., Beightol S., Mandahar U., Suzuki R., Xiao S., Lu H. W. and Novak J. T. (2017). Pretreatment of a primary and secondary sludge blend at different thermal hydrolysis temperatures: impacts on anaerobic digestion, dewatering and filtrate characteristics. *Water Research*, **122**, 557–569.

Hollowed M., Chan A. A., Zhao H., Christensson M., Lemaire R. and Viswanathan S. (2019). Implementation of IFAS ANITA Mox Deammonification Process. How Sundet WWTP converted existing MBBR to tackle high strength thermal hydrolysis sidestream. Proceedings of Water Environment Federation's WEFTEC Conference, Chicago, IL, USA.

Inkpin L., Merry J., Madden G. and Fountain P. (2019). Successful commissioning of Basingstoke STW Demon® liquor treatment plant. Proceedings European Biosolids and Organic Resources Conference, 19th–20th November 2019, Manchester, UK.

Jolly M., Knight G. and Lloyd A. (2012). Commissioning of Davyhulme Thermal Hydrolysis Plant. 17th European Biosolids and Organic Resources Conference, Aqua Enviro, Leeds.

Kanders L. and Sargalski W. (2014). NYE Mjosanlegget, Lillehammer in Norway, Biogas Plant with THP for Organic Household Waste. Proceedings of Aqua-enviro, 19th European Biosolids and Biowaste Conference, Manchester, UK.

Kartal B., de Almeida N. M., Maalcke W. J., Op den Camp H. J., Jetten M. S. and Keltjens J. T. (2013). How to make a living from anaerobic ammonium oxidation. *FEMS Microbiology Reviews*, **37**(3), 428–461.

Killilea J. E., Colleran E. and Scahill C. (2000). Establishing procedures for design, operation and maintenance of sewage sludge anaerobic treatment plants. *Water Science and Technology*, **41**(3), 305–312.

Lancaster R. (2015). Thermal Hydrolysis at Davyhulme WWTW, one year on. Water Environment Federation (WEF) Residuals and Biosolids.

Li Y. Y. and Noike T. (1992). Upgrading of anaerobic digestion of waste activated sludge by thermal pretreatment. *Water Science and Technology*, **26**(3–4), 857–866.

Li T., Martin Ruel S., Garcia-Belinchón C., Mehta S. and Dauthuille P. (2016). Characterization of anaerobic digestion return liquor, IWA Singapore International Water Week.

Liu X., Wang W., Gao X., Zhou Y. and Shen R. (2012). Effect of thermal pretreatment on the physical and chemical properties of municipal biomass waste. *Waste Management*, **32**(2), 249–255.

Loomis P., Guven E. and deBarbadillo C. (2015). Starting from Scratch – Commissioning the First Thermal Hydrolysis Fed Digesters in North America. Proceedings of Water Environmental Federation WEFTEC, 3379–3388.

Lu H. W., Xiao S., Le T., Al-Omari A., Higgins M., Boardman G. and Murthy S. (2014). Evaluation of solubilization characteristics of thermal hydrolysis process. *Proceedings of the Water Environment Federation*, **2014**(15), 6312–6336.

Lusk L. T., Goldstein H. and Ryder D. (1995). Independent role of beer proteins, melanoidins and polysaccharides in foam formation. *Journal of the American Society of Brewing Chemists*, **53**(3), 93–103.

Marneri M., Jolis D., Kan H., Ahmad M. L. and Pitt P. (2003). Thermal hydrolysis of secondary scum for control of biological foam. 25th Annual Hawaii Water Environment Association (HWEA) Conference, Hawaii.

McCarty P. L. (1971). Energetics and bacterial growth. *Organic Compounds in Aquatic Environments*, **1**, 157–172.

Molokwu O. and Rus E. (2017). Interactions of cations, phosphates, M/D and EPS. Presented at SludgeTech Conference, London, UK.

Oosterhuis M., Ringoot D., Hendriks A. and Roeleveld P. (2014). Thermal hydrolysis of waste activated sludge at Hengelo wastewater treatment plant, The Netherlands. *Water Science and Technology*, **70**(1), 1–7.

Pathak B., Al-Omari A., Smith S., Passarelli N., Suzuki R., Khakar S. and DeBarbadillo C. (2018). Vivianite Occurrence and Remediation Techniques in Biosolids Pre-treatment Process. *Proceedings of the Water Environment Federation*, **2018**(4), 103–117.

Phothilangka P., Schoen M. A. and Wett B. (2008). Benefits and drawbacks of thermal pre-hydrolysis for operational performance of wastewater treatment plants. *Water Science and Technology*, **58**(8), 1547–1553.

Rittmann B. E. P. and McCarty P. L. (2001). Environmental biotechnology: principles and applications. McGraw Hill, New York, 400.

Román E., Solvang W. D., Jonassen O. and Lysfjord T. (2012). Energy produced from Waste – Experiences from Norway.

Schneider Y., Beier M. and Rosenwinkel K. H. (2011). Determination of the nitrous oxide emission potential of deammonification under anoxic conditions. *Water Environment Research*, **83**(12), 2199–2210.

Schoepp T., Bousek J., Beqaj A., Fiedler C., Wett B., Fuchs W. and Weissenbacher N. (2018). Nitrous oxide emissions of a mesh separated single stage deammonification reactor. *Water Science and Technology*, **78**(11), 2239–2246.

Speece R. E. (2008). Anaerobic Biotechnology and Odor/Corrosion Control for Municipalities and Industries. Archae Press.

Speece R. E. and Parkin G. F. (1983). The response of methane bacteria to toxicity. Proceedings of the 3rd International Symposium on Anaerobic Digestion, Agosto, **14**, 19.

Stronach S. M., Rudd T. and Lester J. N. (1986). Anaerobic Digestion Processes in Industrial Wastewater Treatment. Springer Verlag, Berlin Heidelberg. Chapter 4. Start-Up.

Svennevik O. K., Beck G., Rus E., Westereng B., Higgins M., Solheim O. E. and Horn S. J. (2019a). CNash-A novel parameter predicting cake solids of dewatered digestates. *Water Research*, **158**, 350–358.

Svennevik O. K., Solheim O. E., Beck G., Sørland G. H., Jonassen K. R., Rus E. and Nilsen P. J. (2019b). Effects of post anaerobic digestion thermal hydrolysis on dewaterability and moisture distribution in digestates. *Water Science and Technology*, **80**(7), 1338–1346.

Tampio E., Ervasti S., Paavola T., Heaven S., Banks C. and Rintala J. (2014). Anaerobic digestion of autoclaved and untreated food waste. *Waste Management*, **34**(2), 370–377.

Tattersall J. M., Belshaw D., Knight G. and Jolly M. (2014). Commissioning of the Advanced Digestion Plant at Davyhulme. Proceedings of Water Environmental Federation, WEFTEC Conference, 1262–1280.

Taylor D. (2019). Medina – Commissioning the First Cambi B2 System in North America Presented at Cambi Academy, Florida, USA.

van Veldhoven J.-E. and Smits A. (2019). WWTP Tilburg. Energy and Resource factory. Commissioning THP and Anammox, things you never hear at a conference, Presented at Cambi Academy, Florida, USA.

Wang J.-W., Li W., Jian Q.-H., Han X.-Y., Hao S. and Tang C.-C. (2018). Digestion Enhancement with Thermal Hydrolysis Pretreatment at Beijing's Gaobeidian Water Reclamation Plant – Operating Experience and Challenges. Proceedings of Water Environment Federation WEFTEC Conference, 2821–2834.

Weiland P. and Rozzi A. (1991). The start-up, operation and monitoring of high-rate anaerobic treatment systems: Discusser's report. *Water Science Technology*, **24**, 257–277.

Wilson C. A., Tanneru C. T., Banjade S., Murthy S. N. and Novak J. T. (2011). Anaerobic digestion of raw and thermally hydrolyzed wastewater solids under various operational conditions. *Water Environment Research*, **83**(9), 815–825.

Xue Y., Liu H., Chen S., Dichtl N., Dai X. and Li N. (2015). Effects of thermal hydrolysis on organic matter solubilization and anaerobic digestion of high solid sludge. *Chemical Engineering Journal*, **264**, 174–180.

Zhang Q., Vlaeminck S. E., DeBarbadillo C., Su C., Al-Omari A., Wett B. and De Clippeleir H. (2018). Supernatant organics from anaerobic digestion after thermal hydrolysis cause direct and/or diffusional activity loss for nitritation and anammox. *Water Research*, **143**, 270–281.

Zhuo Y., Han Y., Cheng Y., Peng D. C. and Li Y. Y. (2015). Transformation characteristics of carbon, nitrogen, phosphorus and sulfur during thermal hydrolysis pretreatment of sludge with high solid content. *Huan jing ke xue Huanjing kexue*, **36**(3), 1006–1012.

Chapter 5

Benefits

The Bazalgette process, as applied to London, is a total failure. It involves the utter waste of all the manurial matters in the sewage....

J. W. Slatter, 1888

The above words written about Sir Joseph Bazalgette's (As chief engineer for the Metropolitan Board of Works (today known as Thames Water), Bazalgette proposed the first sewage tunnels in London, and considered by many as the forefather of wastewater treatment in the UK) proposal highlight the disconnect that still largely exists between sludge and wastewater practitioners. Often, decisions are made on wastewater treatment with little or no consideration or understanding of the impacts on the sludge produced and its treatment. Considering the costs associated with sludge treatment, it is surprising that this divide remains as wide as it is today. This chapter looks at drivers influencing the use of thermal hydrolysis and its impact on positive use of biosolids.

5.1 INTRODUCTION

There are numerous financial, regulatory and other drivers in play when it comes to deciding on a biosolids strategy. By changing the properties of sludge, thermal hydrolysis is highly influential on many of these drivers and results in economic and ultimately environmental benefits. Figure 5.1 shows influence of thermal hydrolysis with respect to drivers for its employment. Chapter 6 shows, several case studies are presented which highlight practical examples of these.

Author: William Barber
doi: 10.2166/9781789060287_0137

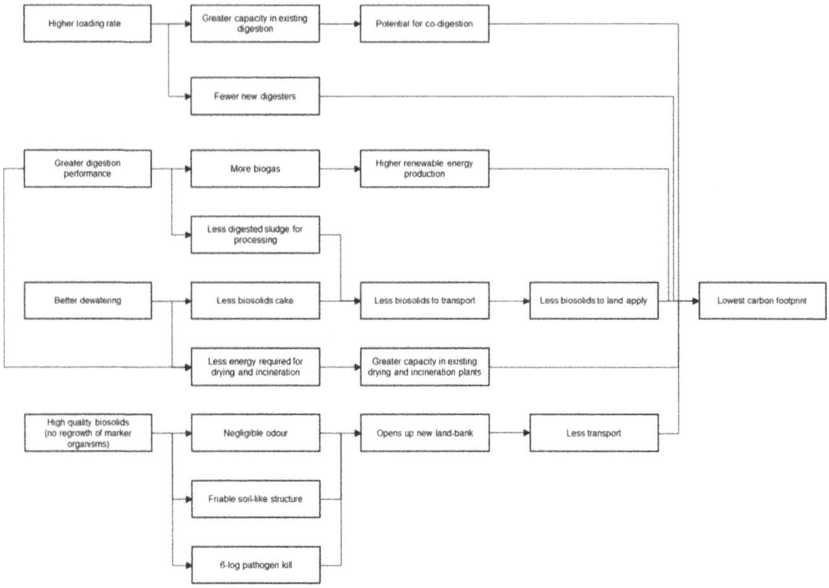

Figure 5.1 Benefits of thermal hydrolysis with respect to sludge processing.

5.1 Higher loading rate in digestion

Conservatively, thermal hydrolysis can double or triple the capacity of existing anaerobic digestion plants

Typically, municipal anaerobic digesters are designed for a loading rate between 2 and 3 kg VS/m^3 digester volume per day, with thickened sludge being fed at a range of 4–6% dry solids (DSs). In addition, retention times are routinely limited to 15 d and longer to accommodate slow kinetics and reduce the potential of biomass washout. As previously described (in Chapter 3) thermal hydrolysis reduces sludge viscosity such that ammonia becomes the limiting step in digestion. Now, feed to a digester can be concentrated to approximately 10% DSs. Furthermore, as the sludge is more biodegradable the retention time is limited by biomass washout rather than kinetics. In this instance, for standard design of municipal digesters where no attempt is made to retain biomass, 10–14 d is sufficient. Typical loading rates for digestion with and without thermal hydrolysis are shown in Table 5.1.

Based on the example of a plant processing 100 t DSs per year, the size of digestion plant required is approximately 32,000 m^3 corresponding to a retention time of 16 d being fed solids at a concentration of 5%. In contrast, installation of thermal hydrolysis would reduce this volume to 13,000 m^3, a reduction of nearly 60%. Building new anaerobic digestion capacity costs

Table 5.1 Typical loading rates for municipal sludge digestion.

Configuration	Loading Rate (kg VS/m^3 d)
Standard anaerobic digestion	2–3
Anaerobic digestion preceded by thermal hydrolysis	5–8

anywhere from 230 to 900 euro/m^3 depending on the region. As thermal hydrolysis reduces digestion capacity by a factor between 2 and 3, it is clear to see that there is a large cost reduction possible. For typical digestion, a volume of approximately 400 m^3/t raw sludge DS is required, and this decreases to 120–140 m^3/t raw sludge DS is required when thermal hydrolysis is used.

Alternatively, the capacity realized on existing digestion plants can be consumed by adding other waste materials for co-digestion. As well as increased biogas production by addition of further load, gate-fees could be charged for receipt of the material. However, the economics of co-digestion are complex and dependent on numerous factors including: material type; level of treatment required; regulations; additional liquor and sludge processing requirements.

5.2 Greater digestion performance

It is well understood that thermal hydrolysis improves the performance of anaerobic digestion. This results in more biogas being produced, and less solids exiting the digester. Table 5.2 shows expected increases in biogas yields for municipal sewage sludge for various configurations of thermal hydrolysis compared to standard mesophilic anaerobic digestion. The range in data is due to differences in sludge composition and operating conditions in the digestion plant.

Table 5.2 Typical biogas production for municipal sewage sludge with different configurations of thermal hydrolysis.

Configuration	Typical Gas Production* (m^3 Biogas/t Dry Solids Fed)
Standard anaerobic digestion	250–360
Upstream thermal hydrolysis of all sludge	400–450
Upstream thermal hydrolysis of only waste-activated sludge	390–430
Intermediate thermal hydrolysis	460–500
Downstream thermal hydrolysis	515–550

*Based on mixed primary and waste-activated sludge

After cleaning to remove siloxanes, hydrogen sulfide and other trace contaminants, this biogas can be exploited in various ways including direct use in boilers, burning in combined heat and power systems, or after purification to remove carbon dioxide, used as a substitute for natural gas on gas grids or as vehicle fuel. The value and end use of the gas produced is highly influenced by locally driven market economics.

5.3 Better dewaterability

Improved performance during digestion coupled with enhanced dewatering results in significantly less biosolids being produced. As shown in Table 5.3, sludge volumes reduce by almost half compared to standard anaerobic digestion. This increases to approximately two thirds for cases where digestion is absent.

Table 5.3 Typical dewatered cake production for municipal sewage sludge with different configurations of thermal hydrolysis.

Configuration	Typical Cake Production* (Wet t/t Dry Solids Fed)
Standard anaerobic digestion	2.5–3.0
Upstream thermal hydrolysis of all sludge	1.6–1.8
Upstream thermal hydrolysis of only waste-activated sludge**	1.8–2.0
Intermediate thermal hydrolysis**	1.4–1.6
Downstream thermal hydrolysis**	0.9–1.2

*Based on mixed primary and waste-activated sludge

The benefits of cake reduction are especially evident in markets where reuse options for biosolids cake are expensive. Additional benefits are attained if further biosolids processing occurs downstream of digestion and dewatering. By producing less cake with less water in it, existing downstream thermal systems such as dryers and incinerators can gain spare capacity in much the same way as anaerobic digestion itself when preceded by thermal hydrolysis. Alternatively, size of such equipment can be significantly reduced on new-build construction projects. Table 5.4 shows the influence of thermal hydrolysis on downstream drying requirements based on sludge being dried to 90% solids.

According to Spalding and Smoot (2019) there are nine thermal hydrolysis plants which have downstream drying, of which five are belt dryers and the remainder are rotary drum machines. Five of the facilities treat mixed sludge and the remainder only activated sewage. The authors suggested that bulk density of thermally hydrolyzed digested dried biosolids is lower than that when thermal hydrolysis is absent, although no data were presented.

Table 5.4 Typical water evaporation required to dry dewatered biosolids cake to 90% DSs in a thermal dryer, based on 100 t DSs per day input to digestion.

Configuration	Water Evaporation* (t Water Evaporation/hr)
Standard anaerobic digestion	7.6–10.0
Upstream thermal hydrolysis of all sludge	4.6–5.3
Upstream thermal hydrolysis of only waste-activated sludge	5.0–5.8
Intermediate thermal hydrolysis	3.7–4.3
Downstream thermal hydrolysis	1.7–4.6

*Based on mixed primary and waste-activated sludge

Due to a combination of improved digestion performance and dewaterability, thermal hydrolysis can significantly increase capacity of downstream drying and incineration facilities

The water evaporation required is indicative of the size of drying plant needed. Depending on the type of dryer and quantity of heat recycled, drying plants consume between 800 and 1,200 kWhr of energy per tonne of water evaporation required. Based on the calorific value of methane this is between 75 and 220 m^3 methane per tonne of water evaporated. This energy can be provided directly by methane, or by using biogas. However, the use of biogas in this way diverts it away from other more beneficial uses such as the production of renewable energy or vehicle use. We can see the influence of thermal hydrolysis by means of an example.

Question. *A plant processing 50 dry t/d of sludge has an existing digestion and drying plant. For mixed sludge, how much energy is left in the form of methane, if biogas is used in the dryer? What is the impact of upstream thermal hydrolysis on both primary and activated sludge on the quantity of surplus energy assuming 1000 kWhr of energy are required per tonne water evaporation and biogas contains 65% methane by volume? How much excess energy is left by use of thermal hydrolysis after accounting for that required by steam generation if sludge is thickened to 16.5% DSs with a temperature difference of 75°C, and system efficiency is 65%. Assume co-generation is available and can account for 90% of this energy demand. Calorific value of methane is 11 kWhr/m^3. Assume digested sludge is dried to 90% solids content. Use information in Tables 5.2, 5.3 and 5.4.*

Answer.

From Table 5.2, taking mean values from the ranges given, standard digestion will produce 360 m^3 biogas per tonne fed to the digester, compared to upstream thermal hydrolysis which is expected to yield 425 m^3/t DSs processed.

The energy content of the biogas produced during standard digestion is calculated as follows:

$= 50 \text{ t DSs/d} \times 360 \text{ m}^3\text{/biogas/t DSs} \times 65\% \text{ methane} \times 11 \text{ kWhr/m}^3 \text{ methane}$

$= 128{,}700 \text{ kWhr/d}.$

By a similar calculation, the energy in the gas produced by thermal hydrolysis is = 151,938 kWhr/d.

To determine the energy required by drying to 90% it is necessary to determine the water evaporation requirement. Taking the average values from Table 5.4, standard digestion with dewatering will require 8.8 t water evaporation/hr/100 t DSs, and thermal hydrolysis will require 4.95 t water evaporation/hr/100 t DSs.

Therefore, for 50 t DSs per day, the daily water evaporation for standard digestion is

$= 8.8 \text{ t/hr} \times 24 \text{ hr/d} \times (50/100) \text{ correction factor}$

$= 105.6 \text{ t water evaporation per day.}$

Similarly, for the option with thermal hydrolysis the daily water evaporation required

$= 59.4 \text{ t water evaporation.}$

The water requires 1,000 kWhr/t water evaporation, therefore for the case when thermal hydrolysis is absent:

$$\begin{aligned}\text{Energy required by the dryer} &= 1{,}000 \text{ kWhr/t water evaporation} \\ &\quad \times 105.6 \text{ t water evaporation/d} \\ &= 105{,}600 \text{ kWhr.}\end{aligned}$$

The same calculation for thermal hydrolysis yields an energy demand for drying of = 59,400 kWhr/d.

Therefore, the surplus energy available for standard digestion = 128,700 kWhr/d − 105,600 kWhr/d = 23,100 kWhr/d.

In this example over 80% of the energy produced by the digestion process is required to fuel the dryer.

For thermal hydrolysis, = 151,938 − 59,400 kWhr/d = 92,538 kWhr. An increase in biogas production coupled with a decrease in energy required by the dryer results in only 40% of the biogas energy produced being needed to maintain the requirements of the dryer when thermal hydrolysis is present.

However, thermal hydrolysis has an energy demand for steam. The conditions are identical to the example given in Section 2.13, therefore, the energy required for steam generation is 2,724,898 kJ/dry t processed.

Therefore, daily energy demand for steam = 50 t/d × 2,724,898 kJ × 1/3,600 (to convert to kWhr) = 37,846 kWhr.

Co-generation plant can account for 90% of this energy demand through the use of high-grade heat, therefore, the energy available from high-grade heat

= 37,846 kWhr/d × 90% = 34,061 kWhr/d.

Therefore, the additional energy required for steam generation is

= 37,846 kWhr/d − 34,061 kWhr/d = 3,785 kWhr/d.

The excess energy available from the employment of thermal hydrolysis after accounting for steam generation can be calculated thus:

= 92,538 − 3,785 kWhr/d = 88,753 kWhr/d.

Although the biogas energy is only approximately 20% higher in the presence of thermal hydrolysis compared to its absence, nearly 4 times as much surplus energy is available for other uses even after the energy demand for steam has been met. This results in a drying system which is self-sufficient for energy while some energy is available for other uses. The benefits further improve with other configurations of thermal hydrolysis (as discussed in Chapter 2, Section 2.4).

5.3.1 Influence of digestion and thermal hydrolysis on incineration

As with drying, incineration is fundamentally influenced by the presence of thermal hydrolysis combined to digestion upstream. This is due to changes in quantity of material sent to the incinerator, changes in percentage of volatile solids, altered composition of the volatile solids, and finally, different dewaterability characteristics. These differences influence the calorific value which in turn controls how much material can be burnt in the incinerator, and how much auxiliary fuel is required if necessary.

Gross calorific value is experimentally measured using bomb calorimetry (British Standard 1016-105). With calorimetry, a known weight of material is ignited and burnt in an insulated vessel, which itself is contained inside a larger vessel containing water in contact with a thermometer. This apparatus is known as a bomb. Thermodynamic properties are calculated based on temperature change recorded. Bomb calorimeters are built to withstand large pressure changes. Electrical energy is used to light the material via a wire connected to the sample, and as the fuel is burning, it heats up the surrounding air, which expands and escapes through a tube in the vessel into the outer vessel containing the water. The temperature of the water allows for calculating calorie content of the fuel. Calorimeters are generally calibrated using a known mass of a highly pure reference standard which is generally benzoic acid.

Alternatively, gross calorific value may be determined theoretically from knowledge of its chemical composition using an empirically derived equation, generally known as the Du Long equation (Technical Report, CEN/TR 13767, 2004). The equation has been developed by many workers since its introduction and exists in many guises (mainly dependent on the units used). The following version is recommended as a first approximation for calculation of gross calorific value of sewage sludge (Technical Report, CEN/TR 13767, 2004):

$$GCV = 32810C + 142246(H - O/8) + 9273S \tag{5.1}$$

where GCV units are in kJ/kg loss on ignition (LOI); C, H, O and S are percentage compositions of carbon, hydrogen, oxygen and sulfur, respectively. Equation (5.1) slightly overestimates the heat value of sludge with high organic nitrogen content due to: nitrogen association with hydrogen as an amine; and the production of nitrogen oxide during amine combustion reduces the heat released from hydrogen. Therefore, the following equation has been recommended to calculate the GCV of sewage sludge:

$$GCV = 32810C + 142246(H - O/8) + 9273S - [2189N(1 - \mu) + 6489N\mu] \tag{5.2}$$

where N refers to the percentage fraction of nitrogen; and μ refers to the mass fraction of nitrogen which is converted to nitrogen oxide (generally in the range of 2–7%). Using these equations with typical composition of various sludges, energy content of sludges are approximately, 25,700 and 21,800 kJ/kg volatile material for primary and typical activated sludge, respectively.

During digestion, volatile material within the sludge is converted to biogas with the remainder continuing within the digested sludge which exits the digester. The biogas removes carbon in the form of both methane and carbon dioxide, hydrogen in the form of methane (and minuscule quantities as hydrogen gas by comparison), and trace quantities of nitrogen (in ammonia and similar compounds) and other minor compounds. The extraction of these elements changes the composition of what remains within the volatile fraction of the sludge. As inferred in equations 5.1 and 5.2, carbon, hydrogen and sulfur contribute to energy content while oxygen and nitrogen deplete it. By combining knowledge of the sludge elemental composition with a simple energy balance around digestion, it is possible to calculate the influence of volatile solids destruction – ultimately thermal hydrolysis – on elemental composition within the volatile sludge fraction. This is shown in Figure 5.2:

Figure 5.2 shows, ironically, that the greater the volatile solids destruction is, and the more carbon, hydrogen and nitrogen are lost to the biogas, the greater the carbon content in the remaining volatile solids fraction. This is because more oxygen is lost in carbon dioxide, than carbon in both carbon dioxide and methane and hydrogen in methane.

> *Calorific value of thermally hydrolyzed digested biosolids cake can be similar or even higher than raw cake due to improved dewaterability*

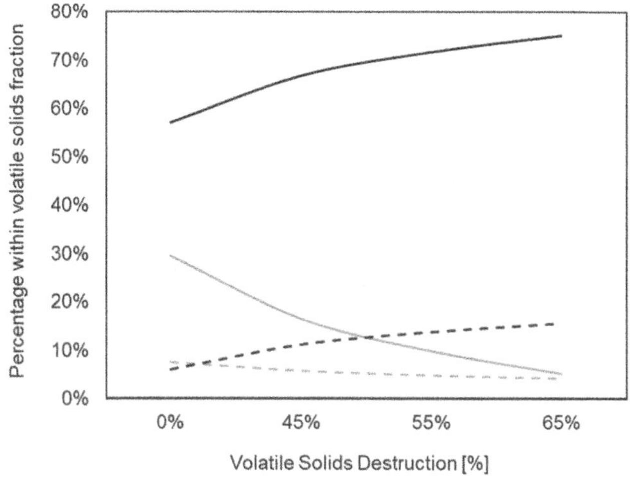

Figure 5.2 Influence of volatile solids destruction on changes in elemental composition of remaining volatile solids fraction of digested sludge. Key: black line = carbon; gray line = oxygen; black dashed line = nitrogen; gray dashed line = hydrogen.

Therefore, oxygen is degraded in the remaining volatile material at a faster rate than other elements, which increase. Although hydrogen also reduces in the volatile solids fraction due to improved digestion, its quantity relative to carbon and oxygen is small and therefore less influential on overall energy content. The consequence of this being that calculations of energy content of digested sludge based solely on ash content underpredict the energy held within the sludge.

The following correction factor has been derived from the data analysis above to describe the influence of volatile solids destruction on energy content within the volatile solids fraction remaining:

$$CV_{VS\ fraction} = 8{,}007\ V + 23{,}800 \tag{5.3}$$

where $CV_{VS\ fraction}$ is the energy content of the volatile solids fraction in kJ/kg; and V is the volatile solids destruction expressed as a decimal. Its use is best explained by means of an example.

Question. *Raw sludge has a calorific value of 18,500 kJ/kg dry substance. It contains 75% volatile solids. The sludge is thermally hydrolyzed and digested such that the digested sludge now only contains 55% volatile fraction. Thermal hydrolysis and digestion results in a 60% destruction in the volatile solids. Determine the drop in calorific value assuming (a) that there is no change in elemental composition in the volatile solids fraction, and (b) that it changes and is described by equation 5.3. What is the percentage difference in energy content between the two methods?*

Answer.
If it is assumed that there is no change in volatile solids composition, then the calorific value can be determined pro-rata by comparing the volatile solids content, as follows:

Energy content $= 18{,}500 \times (55/75) = 13{,}567$ kJ/kg

Now digestion alters the elemental composition. Therefore, the energy content in the volatile solids fraction is:

$= 8{,}007 \times 0.6$ (volatile solids destruction) $+ 23{,}800 = 28{,}604$ kJ/kg

Volatile solids content is 55% of the sludge, therefore, the energy content is $0.55 \times 28{,}604 = 15{,}732$ kJ/kg.

This is an increase of 16% compared to the case when no change in elemental composition is accounted for. On large incinerators this is a significant difference.

The greatest impact of thermal hydrolysis and digestion on incineration, however, is its impact on dewatering. While the energy content of the sludge reduces on a dry basis, changes in dewaterability make this reduction less clear. Improved dewaterability entails that the energy within the sludge is diluted less when compared to a case where dewaterability is worse. If one looks at the previous example and assumes dewaterability of 25% and 32% for raw and processed sludge cakes, respectively, then the calorific values become $18{,}500 \times 0.25 = 4{,}625$ and $15{,}752 \times 0.32 = 5{,}040$ kJ/kg wet cake for raw and thermally hydrolyzed digested materials, respectively. While on a dry basis the processed sludge had a calorific value nearly a third lower, when compared as a wet material the energy contents are similar, if not slightly higher for the material which has been hydrolyzed and digested. This is key in attaining spare capacity in existing incineration plants.

If, as is the case with this example and additional processing has increased the energy content on a cake basis, spare capacity can be derived in two ways. First, due to a reduction in sludge mass because of improved digestion and conversion to biogas upstream, and secondly by not requiring as much auxiliary fuel or diversion of heat at the incinerator. If digestion performance but not dewaterability is improved, the cake has a lower energy content for the digested option. Therefore, any potential capacity increase by improved digestion has to be offset by requirement of additional auxiliary fuel, such as natural gas at the incinerator. This was a key factor in the employment of thermal hydrolysis at Davyhulme which will be discussed later. The influence of volatile solids destruction combined with dewaterability on calorific value is shown in Figure 5.3.

The lines in Figure 5.3 can all be described to derive an empirical relationship based on a second-order polynomial of the form:

$$Y = aV^2 + Vx + c \tag{5.4}$$

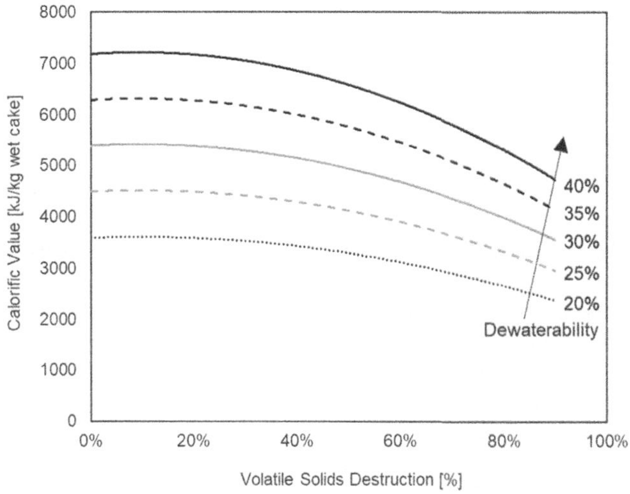

Figure 5.3 Influence of volatile solids destruction and dewaterability on calorific value of sludge cake. Baseline raw sludge is considered to contain 75% volatile solids content.

Constants *a*, *b* and *c* are dependent on dewaterability performance and described in Table 5.5 and *V* is the volatile solids destruction expressed as a decimal as previously discussed.

Question. *What is the difference in calorific value of raw cake which dewaters to 25% compared to thermally hydrolyzed digested cake which dewaters to 30%? For thermally hydrolyzed cake, assume 60% volatile solids destruction. How do these compare to digested cake where volatile solids is 55%, but dewaterability does not improve and remains at 25% DSs?*

Answer.
At 25% cake solids, from Table 5.5 constants for use in equation (5.4), $a = -2{,}384.6$, $b = 445$ and $c = 4{,}490$

From equation (5.4):

Calorific value $= -2{,}384.6\ V^2 + 445\ V + 4{,}490$

Table 5.5 Constants for use in equation 5.4.

Dewaterability (% Dry Solids)	*a*	*b*	*c*
20	−1907.7	356	3592
25	−2384.6	445	4490
30	−2861.5	534	5388
35	−3338.5	623	6286
40	−3815.4	712	7184

In this case, volatile solids destruction $V = 0$ as there is no digestion, and calorific value becomes 4,490 kJ/kg wet cake.

For thermally hydrolyzed digested cake with 30% DSs, from Table 5.5, $a = -2{,}861.5$, $b = 534$ and $c = 5{,}388$, $V = 0.6$ (60% volatile solids destruction).

Substituting into equation (5.4):

$$\text{Calorific value} = -2{,}384.6\ V^2 + 445\ V + 5{,}388 = 4{,}796\ \text{kJ/kg wet cake.}$$

This is 306 kJ (7%) higher compared to the raw cake. Although volatile solids are lower, enhanced dewaterability results in a comparable energy content between the raw and enhanced digested cake. If dewaterability is not similar or worse, then calorific value of the digested cake will reduce when benchmarked against raw sludge. This is a key consideration when determining the potential spare capacity in incineration downstream of enhanced and standard digestion.

Now to compare with digestion at 55% volatile solids destruction but no improvement on dewatering.

As for the raw sludge,

$$\text{Calorific value} = -2{,}384.6\ V^2 + 445\ V + 4{,}490$$

$$\text{Substitute } V = 0.55$$

$$\text{Calorific value} = 4{,}013\ \text{kJ/kg.}$$

As dewatering is not improved, calorific value drops accordingly due to the destruction of volatile solids during digestion. In this instance, enhanced capacity due to reduced sludge input is partly offset by the sludge having lower energy content. An example of this is given later, in the case studies.

5.4 Higher quality biosolids product

5.4.1 Biosolids cake

By sterilizing sludge to greater than the temperatures required for pasteurization, thermal hydrolysis, when applied to all sludge flow, produces a biosolids material free of pathogenic marker organisms. This results in the biosolids meeting guidelines for advanced treated materials, such as those found in the UK's Safe Sludge Matrix, or the US EPA's 503 regulations. Typically, advanced treated materials are those which kill 99.9999% (6-log) of pathogens and are free of *Salmonella*. In the US EPA regulations advanced treated biosolids is designated as 'Class A' which is defined as sludge which has a density of fecal coliforms in the sludge of less than 1,000 most probable number (MPN) per g of total DSs, or a density of *Salmonella* sp. bacteria lower than 3 MPN/4 g of total DSs, and meets a time/temperature relationship according to various equations.

There are numerous benefits of treating biosolids to this standard, most important of which being less onerous permitting requirements and the ability to use the biosolids on a wider range of land, as shown in Table 5.6, adapted from the UK's

Table 5.6 Influence of sludge treatment type on type of land available for sludge reuse based on the Safe Sludge Matrix.

Crop Group	Untreated Sludge	Conventionally Treated Sludge	Advanced Treated Sludge
Fruit	Prohibited	Prohibited	Allowed[3]
Salads	Prohibited	Prohibited[1]	Allowed[3]
Vegetables	Prohibited	Prohibited[2]	Allowed[3]
Horticulture	Prohibited	Prohibited	Allowed[3]
Combinable and animal feed crops	Prohibited	Allowed	Allowed
Grass and forage (grazing)	Prohibited	Prohibited[4, 5]	Allowed[5]
Grass and forage (harvesting)	Prohibited	Allowed[4, 6]	Allowed[5]

[1]30-month harvest interval applies
[2]12-month harvest interval applies
[3]10-month harvest interval applies
[4]Deep injected or ploughed only
[5]3 week no razing and harvest interval applies
[6]No grazing in season of application

Safe Sludge Matrix. Figure 5.4 shows a typical thermally hydrolyzed digested cake product.

However, other than meeting the requirements for 'advanced' or 'Class A' standard, there are numerous other variables which influence how desirable a biosolids material is. One of the most important is the odor potential of the material. Several studies were set up to elucidate the reasons behind odor generation and observed sudden increases in fecal coliform densities immediately after sludge dewatering (Higgins *et al.*, 2008; Murthy *et al.*, 2009). This work, known in the industry as ROSI (Regrowth, odors and Sudden Increase) studied a variety of sludge treatment and dewatering configurations, and concluded that sludge processed by thermal hydrolysis and digestion had minimum odor potential based on measurements of volatile organic sulfur compounds especially when dewatered using belt presses. Figure 5.5 is adapted from the findings of the study.

Figure 5.5 shows processes which, in theory, are treating sludge in excess of the requirements of the US EPA 503 regulations for Class A biosolids, as previously defined. However, after dewatering, enzymic hydrolysis (similar to acid phase digestion), pre-pasteurization with heating to 70°C, and temperature phased digestion (thermophilic followed by mesophilic) all failed to meet the requirements for marker organisms. Only thermal hydrolysis and long retention time staged thermophilic digestion met the requirements of the regulations.

Figure 5.4 Thermally hydrolyzed digested biosolids from Kapuściska, Poland. Courtesy Cambi.

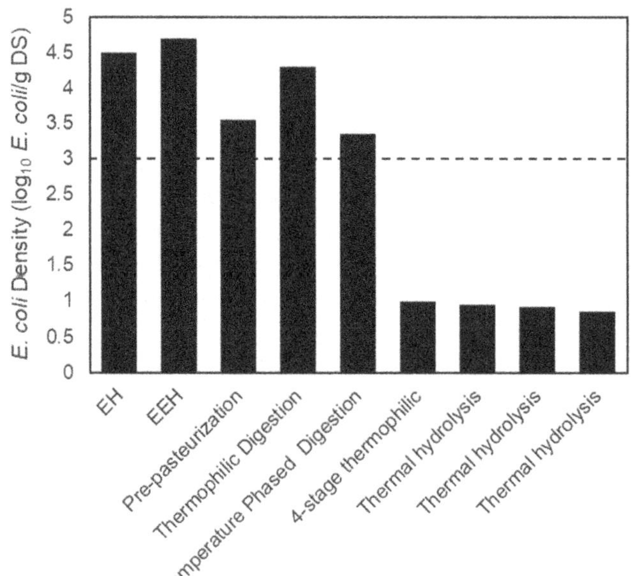

Figure 5.5 Measurements of regrowth for advanced treated biosolids materials measured after centrifuge dewatering. The dotted line shows the limit for achieving Class A biosolids as defined in the US EPA's 503 Regulations. EH and EEH refer to Enzymic and Enhanced Enzymic Hydrolysis, respectively. Adapted from Higgins *et al.* (2008).

Regarding the potential for reactivation of marker organisms, Murthy *et al.* (2009) showed *E. coli* DNA density below a threshold figure of approximately 3–5 × 10^4/g DS for sludge which had been thermally hydrolyzed, above which reactivation was highly probable. This result was much lower than when compared with other technology designed for pathogen destruction including pre-digestion pasteurization and temperature phased digestion.

The study found a close link between regrowth and the odors released. Typically, the greater the regrowth, the greater the odors within the biosolids. A complementary study commissioned by the Water Environment and Research Foundation (WE&RF) investigating the odor potential of a variety of Class A treated biosolids is currently ongoing (Brandt, 2018). One of the key findings to date is a benchmark comparing treatment processes to the odor dilution threshold (DT). The dilution to threshold measure is a ratio of a volume of filtered air to a volume of ambient air. If the sample can be smelt, then the test sample is further diluted. This continues until the odor can no longer be detected (Brower *et al.*, 2017). Therefore, the more dilutions required (higher DT number) the greater the odor. Figure 5.6 has been adapted from the study.

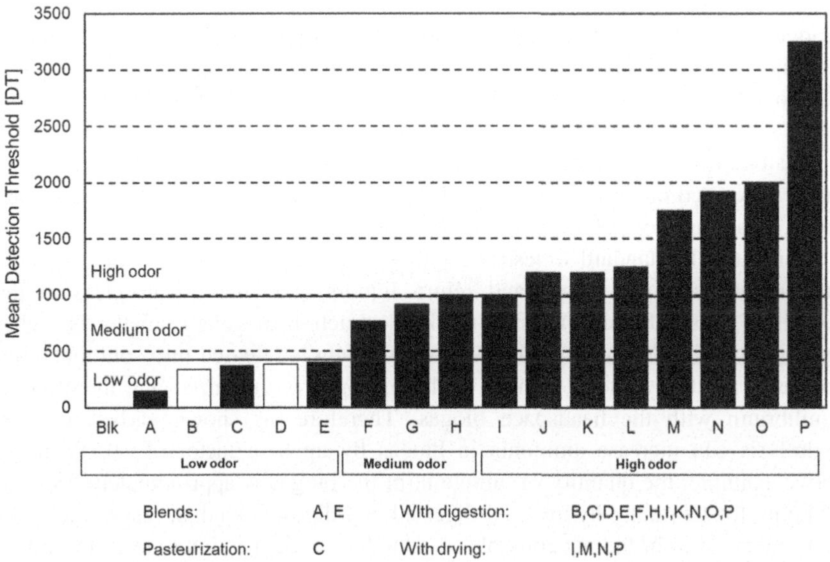

Figure 5.6 Mean detection threshold for odors for a variety of biosolids products. Adapted from data presented on ongoing study (WE&RF Project NTRY7R15 – High Quality Biosolids from Wastewater – led by Material Matters). The higher the value, the greater the odor. Key: white bars – plants with thermal hydrolysis; black bars – other. High-level description of configuration given on chart.

Processes with low DT are typically based on biosolids which have been blended with wood products for an extended time period, and thermal hydrolysis (white bars in Figure 5.6). It was suggested that processes based on variants of digestion had medium odors, and processes based on drying typically had high odors regardless of digestion or not. The study is aiming to determine methods of predicting the DT of biosolids materials (Elliott *et al.*, 2018). Brower *et al.* (2017) presented data on difference in DT or thermally hydrolyzed cake compared to the limed cake they had previously. About 90% of the odors associated with thermally hydrolyzed cake had DT threshold of less than 2 with the remaining 10% under 4. By comparison, limed material had a wider range of odors. Just over half were below 2, and approximately a third under 4. While the remainder of the odor was predominantly within a DT of 4–7, there were trace odors which had values as high as 30 and even 60. In the same study, limed cake odors were described as musty with manure and ammonia notes, and traces of fishy and musky smells. By contrast, some of the musty smell with the thermally hydrolyzed digested cake was replaced by earthy and burnt wood aromas. An ammonia scent was also present but less obvious than for the limed material.

5.4.2 Nutrient content of thermally hydrolyzed sludge cake

Sludge and biosolids contain both nitrogen and phosphorous, both of which can be exploited by applying treated biosolids on agricultural land. As previously mentioned, nitrogen is released during the breakdown of proteinaceous materials in the sludge, and with knowledge of the nitrogen content of the feed material, it is possible to determine a nitrogen balance and determine the quantity remaining in the cake used on agricultural land. This is shown in Figure 5.7. The figures are based on an input of 100 dry t/d into anaerobic digestion.

Compared to standard digestion, hydrolyzed digested sludge results in more nitrogen in both the biogas and return liquors and subsequently less in the biosolids. As solid concentration and the destruction of volatile matter is higher in digesters with thermal hydrolysis more nitrogen is released into solution. With increased alkalinity and pH, more of that nitrogen is unionized and therefore in equilibrium with the headspace biogas. Therefore, as shown earlier, thermal hydrolysis can increase ammonia in biogas by up to a factor of 10. With the above example, the quantity of ammonia in the biogas is approximately 580 and 40 ppmv for thermally hydrolyzed digested and digested sludge, respectively. As more nitrogen is in the aqueous phase, a higher amount is returned in the liquid removed from dewatering. In the example shown in Figure 5.7, this is approximately a third higher. Accounting for higher levels of nitrogen in the return dewatering liquors and biogas, there is approximately 20% less nitrogen in cake produced from thermal hydrolysis and digestion compared with digested cake in the absence of hydrolysis.

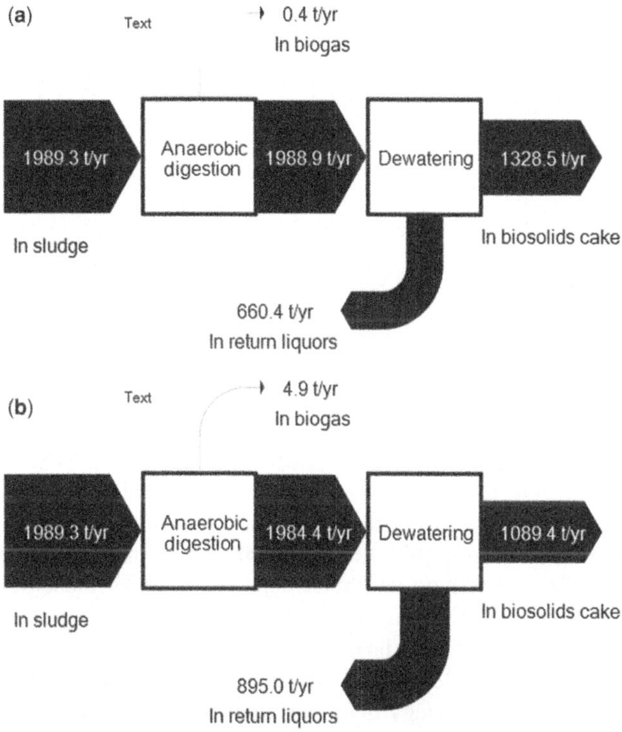

Figure 5.7 Typical nitrogen balance across anaerobic digestion and dewatering. Key: (a) standard anaerobic digestion; (b) digestion preceded by thermal hydrolysis. Example based on processing 100 t DSs per day. Figures are in t/year.

From mass balance, carbon to nitrogen ratio is approximately 7 and this compares well with data presented by Alexander and Peot (2017) with a value of 6.6. This was lower than compared to composted materials which were in the range of 10–12. For standard digestion, a value of approximately 9 was obtained by calculation, which is to be expected as more carbon is converted to biogas in the presence of thermal hydrolysis.

In terms of concentrations, nitrogen content of thermally hydrolyzed cake is calculated as approximately 5.5% and 1.8% on a dry and wet basis respectively. (For comparison the standard digested cake returns values of 5.8% and 1.3%. These calculated values compare well with the literature, as shown in Table 5.7.)

Clearly, the nitrogen and phosphorous content depends on the type of sludge produced, which in turn is reliant on choice of wastewater treatment. As mentioned, mass balance would suggest that nitrogen content of sludge which has been thermally hydrolyzed and digested would be lower than that produced from digestion alone.

Table 5.7 Nitrogen and phosphorous content of thermally hydrolyzed digested cakes.

Sample	DSs (%)	Volatile Solids (%)	pH	Nitrogen (% dry basis)	Nitrogen (% wet basis)	Phosphorous* (% dry basis)	Phosphorous* (% wet basis)
Hias	27	57	8.6	3.5	0.9	2.4	0.6
Lillehammer	26		8.2	8.5	2.2	3.4	0.9
Nigg	32	57	7.9	3.8	1.2	3.3	1.1
Cotton Valley	25–30	55	7–7.5	4.1	1.1	2.4	0.7
Ringsend	27	54	7.1	3.9	1.1	1.9	0.5
Blue Plains	31.5	59	7.8	4.4	1.4	2.8	0.9
Næstved	28			4.8	1.3	3.0	0.8

Adapted from Blytt (2009); data on DC Water Bloom products, and operating data
*To convert to P_2O_5 equivalents multiply by 2.3

Blytt (2009) summarizes the experience of five plant owners with respect to land application of thermally hydrolyzed digested and other biosolids products. There were several key similarities between the owners. Prior to thermal hydrolysis, there was either no or limited land recycling of biosolids. Biosolids volumes were large, material typically had inconsistent quality and composition, and importantly, biosolids would have unpleasant odors. This led to difficulties in land-recycling practices and biosolids storage. Among other factors, these were key reasons to move to a strategy involving thermal hydrolysis. After installation of the technology, uptake and interest in biosolids use increased substantially. A significant reduction in biosolids quantity (see Table 5.3) was realized, and this improved logistics and opened options for off-site farm storage. Besides a reduction in quantity, farm storage became of interest since the material did not absorb rainwater to a significant degree. Other key desirable attributes were the crumbly soil-like nature of the material, and the significant reduction in odor and change in its hedonic tone.

Most of the clients also undertook marketing drives with some developing brand names for the material, such as Anglian Water's *nutri-bio*. Success of marketing was based on proactive open and transparent communications with end users, education and interaction with key stakeholders, employing staff or consultants with experience of farming practices, data collection and product trials. Interestingly, two of the owners, Anglian Water and Celtic Anglian Water (owners of Ringsend facility in Dublin) produced a variety of materials, including dried pellets, to accommodate different land types and provide flexibility. In both instances, farmers, when given a choice, would preferentially opt for biosolids cake over dried material (Blytt, 2009). Reasons for the choice of cake over dried pellets included: higher concentration of organic material; observed faster improvement of fields treated with cake than with granules; cake being better to apply to improve soil than granules; improved soil structure and water storage capacity with sandy soils. Farmers have been very positive with the use of thermally hydrolyzed biosolids cakes (Lancaster, 2015).

Besides nutrients such as nitrogen and phosphorus, sludge-derived cakes contain a variety of metals and xenobiotics. Table 5.8 shows metal concentrations of typical biosolids compared with three sites which employ thermal hydrolysis.

There are four major removal mechanisms for metal removal during anaerobic digestion and these are: adsorption, absorption, precipitation and complexation (Lester, 1987a, 1987b). The factors influencing these removal mechanisms may be divided into three broad groups (Lester, 1987c): (1) operating parameters – sludge age, suspended-solids removal, mixed liquor suspended solids concentrations, solids retention time, hydraulic retention time, dissolved oxygen concentration, settling time and reactor characteristics; (2) physical/chemical parameters – temperature, pH, metal ion concentration, metal solubility, metal valency, concentration of complexing agents and particle size and (3) biological parameters – concentration of extracellular polymers.

Thermal hydrolysis influences these parameters in various ways. Steam explosion reduces particle size, which in turn increases particle surface area and charge. This directly influences the propensity for both adsorption and absorption. Due to a combination of increased loading rate and enhanced destruction of volatile solids, pH during digestion increases by as much as one unit. As protons compete with metal cations for binding sites, potential for chemical precipitation, adsorption and absorption may be reduced. Finally, increased solubility, destruction of extracellular polymers, improved digester stability and production of natural chelating agents such as humic and fulvic acids would be expected to impact on the chelating potential within the digester. However, as can be seen from Table 5.8, there appears no obvious difference between thermally hydrolyzed digested cakes and those produced from digestion alone. This suggests that the initial metal concentrations which are reliant on site-specific parameters are of greater influence than any chemical or biological changes which occur during anaerobic digestion.

5.4.3 Thermally hydrolyzed digested compost

Although not formally required to meet existing biosolids guidelines for agricultural use of biosolids, pioneering work has been conducted by DC Water on further improving the properties of cake material by additional processing and composting (Alexander & Peot, 2017; Brower *et al.*, 2018). This work has looked at addition of various green wastes, curing (such as simple and windrow composting) and scalping – defined as thermal drying to increase DSs by several percent but cutting well short of producing a dried pellet. The influence of storage on both dry and volatile solids content is demonstrated in Figure 5.8.

As expected and shown in Figure 5.8, the DSs increases and volatile solids is reduced by storage. An improvement of approximately 20 percentage points is

Table 5.8 Typical metal concentrations in sludge with and without thermal hydrolysis, compared with regulatory limits.

	Standard** Digested	Typical EU***	Typical USA***	Blue Plains	Nigg	Næstved	Gaobeidian****	US EPA	UK	Europe Range***
	No thermal hydrolysis			With thermal hydrolysis						
Aluminum*				7,900						
Arsenic	25			4.7			8–29	41	50	
Cadmium	1.6	2.8	16	1.0	0.8	2.0	<2.5	39	3	20–40
Chromium	62	141	890	49	14.0	28	54–144	1,200	400	
Molybdenum				13	<10			75	4	
Lead	210	124	500	27		96	10–41	300	300	750–1,200
Mercury	1.3	2.2	5	<1	0.3	1.5	4–14.6	17	1.0	16–25
Copper	335	337	850	390	20	330	100–415	1,500	80	1,000–1,750
Nickel	69	37	82	24		25	11–61	420	50	300–400
Manganese	361	260		300						
Cobalt	4.3			6.0						
Iron				76,000						
Selenium				4.3				36	3	
Zinc		1,222	1,740	710		1,828	630–1,790	2,800	200	2,500–4,000
Fluoride					33				500	
Potassium					675					

*Concentration of all materials in mg/kg
**Average based on several samples of digested sludge
***Taken from Inglezakis *et al.* (2014). Treatment type for typical EU and USA sludges unspecified
****Taken from Wang *et al.* (2018)

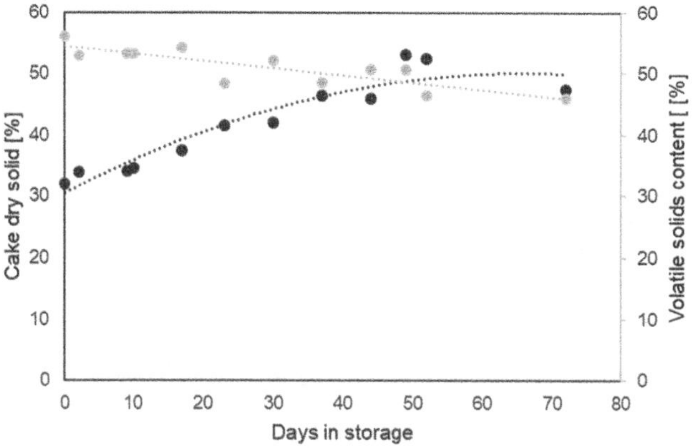

Figure 5.8 Influence of storage on cake DSs content (primary axis – black data) and volatile solids content (secondary axis – gray data) of thermally hydrolyzed cake. Plot using data presented by Alexander and Peot (2017).

observed in the cake within 50 d, after which there appears no further increase. Volatile solids decreases in a linear fashion.

In terms of bulk density there appeared little difference between fresh biosolids cake and cured material, with both being approximately 730 kg/m^3. However, these data were higher for composted biosolids blends with DSs above 70%. These were in the range of 450–600 kg/m^3.

Table 5.9 compares fresh cake with cured material after 1 month. Data on biosolids composts is presented for comparison.

It is difficult to compare the information on biosolids compositing in the absence of operating conditions and type and quantity of green waste added. Needless to say, C:N ratios are higher in those materials, as would be expected. There appears to be little other differences in the material blends.

5.5 Carbon footprint reduction

The various benefits of thermal hydrolysis combine to significantly reduce the carbon footprint of biosolids processing regardless of end use of the biosolids (Barber, 2009a). Subsequently, various independent studies looking at carbon impacts of biosolids process options have concluded that the installation of thermal hydrolysis results in lowest potential carbon footprint when benchmarked against options (Asian Development Bank, 2012; Auty & Blake, 2009). This appears to be due to several reasons as mentioned below:

- Increased production of renewable energy
- Better volatile solids destruction resulting in less biosolids downstream for transport and further processing

Table 5.9 Comparing fresh thermally hydrolyzed cake with thermally hydrolyzed cake after 30 d storage and other biosolids composts.

		Fresh Cake* (Bloom)	**Cured Bloom****	**Biosolids compost 1*****	**Biosolids compost 2*****
DSs	(% DS)	31	42	70	69
Volatile solids	(% VS)	56.1	50.9	53.4	64.3
pH		6.6	6.8	8.3	8
C:N ratio		7.0	6.9	9.8	12
Arsenic	(mg/kg dry)	2.9	11	11	13
Cadmium	(mg/kg dry)	0.8	3.5	1.2	<1.0
Copper	(mg/kg dry)	250	430	180	210
Lead	(mg/kg dry)	21	49	20	18
Mercury	(mg/kg dry)	0.3	<1.0	<1.0	<1.0
Molybdenum	(mg/kg dry)	9	11	7.8	3.2
Nickel	(mg/kg dry)	15	26	16	11
Selenium	(mg/kg dry)	2	3.9	3.3	2
Zinc	(mg/kg dry)	488	720	580	330

*Average between Alexander and Peot (2017) and Bloom analysis from website
**After turning, 30 d
***Data on type, length and quantity of green waste added not provided.

- Better dewatering which further reduces biosolids for downstream processing. Also, significant reduction in fossil fuel requirements for downstream drying
- Higher dewaterability increases energy content in cake which provides greater energy recovery benefit in downstream incineration, while improved volatile solids destruction reduces the quantity of material which needs to be incinerated
- Higher standard of treatment for biosolids means more landbank is opened up which reduces transport distance for biosolids
- Higher loading rates in digestion so less material used in construction which reduces embodied carbon impact – although this is mainly offset by use of stainless steel in reactor vessel construction

In a previous study (Barber, 2009b), new carbon footprint models were developed to determine the carbon footprint of a proposed biosolids management plan (called Sludge Balanced Asset Programme – SBAP). This was benchmarked against the existing scenario of liming at seven sites, and a suggested new incineration plant for raw sludge. The background of this project is described in Chapter 6 Davyhulme case-study. The study looked both at embodied (i.e., those associated with construction) and operating carbon footprints. Compared to a new incineration plant, the SBAP solution reduced embodied carbon from approximately 18,200 to 14,000 t CO_2 equivalents. Although the SBAP would require more concrete to construct, it used approximately a third of the steel that a new incinerator would have required, and it was this parameter which resulted in the drop in embodied carbon footprint. From the paper (Barber, 2009b), carbon footprints were 0.26 t CO_2/t concrete and 6.15 t CO_2/t steel constructed. The interested reader is referred to Hammond and Jones (2008) for additional carbon footprints associated with construction materials.

The operating carbon footprint of the existing scenario was just under 30,000 t CO_2 equivalents annually of which approximately 30% was biosolids transport, a further 30% equally divided by dewatering (with associated liquor treatment) and liming, and the remaining 40% due to land emissions from biosolids application to land. Determination of carbon footprint for SBAP comprised of four separate calculations. First, carbon footprint of transport of sludge from the feeder sites to Davyhulme (where thermal hydrolysis plant was to be installed); second, change in operating carbon footprint at Davyhulme; third, the additional carbon footprint generated at an existing incineration plant to accommodate the additional biosolids generated (and pumped as liquid) from Davyhulme and, finally, the carbon footprint attributed to recycling a fraction of the biosolids produced at Davyhulme to land (Barber, 2009b). Figure 5.9 summarizes these calculations, simplified from Barber (2009b).

Compared to the existing scenario, the carbon footprint was reduced by approximately 32,000 t/year, and this was equivalent to an 8% reduction in carbon footprint in the carbon footprint of the facilities' owner, United Utilities'.

In a later work, a study was conducted to determine the influence of new thermal hydrolysis infrastructure at DC Water's Blue Plains facility compared to a baseline prior to the upgrades (Willis *et al.*, 2018). The overall carbon footprint was reduced from 165,350 down to 115,185 t CO_2 equivalents/year, a reduction of 30%. Scope 1 emissions increased after the upgrade from approximately 41,000 to 51,500 t CO_2 equivalents/year. This was mainly due to an increase in natural gas consumption for steam production, and greater usage of methanol to accommodate increased release of nutrients. Based on previous work by the same team, the impact of methanol addition would have been greater by a factor of 4 but for the installation of a deammonification plant. Scope 2 emissions concerning electricity consumption reduced by circa 30,000 t CO_2 due to less mixing during digestion and other reductions in processing requirements. Finally, Scope 3 emissions reduced mainly due to less transport fuel (10,500 reduced to 3,400 t CO_2 equivalents/year), cessation of lime use (saving of $<$15,000 t CO_2/year) and improved digestion which reduced methane emissions due to land application (21,000 to 1,000 t CO_2/year Willis *et al.*, 2018). However, nitrous emissions increased by over 6,000 t CO_2/year after the upgrade. Carbon credits were

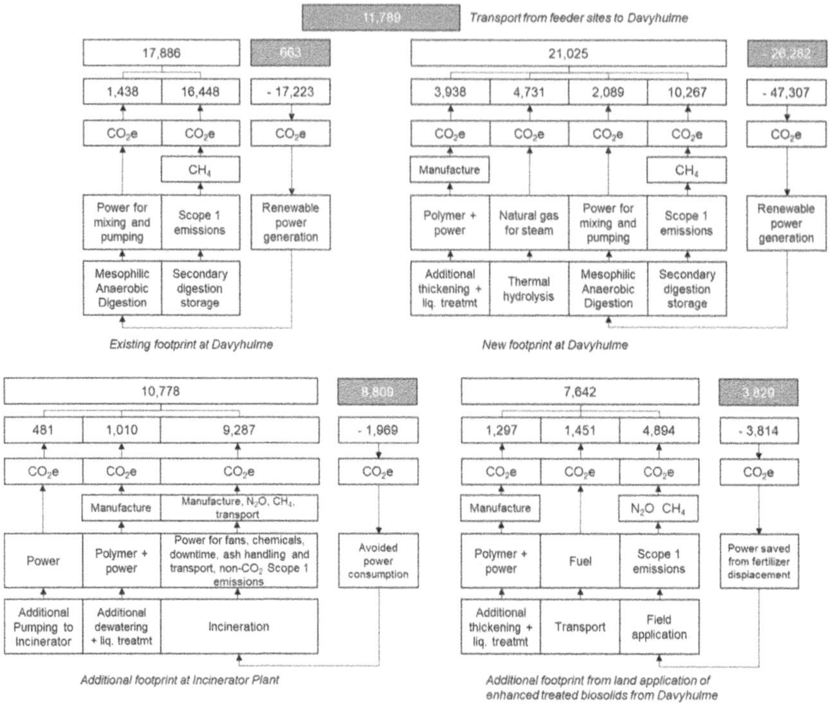

Figure 5.9 Determination of carbon footprint of SBAP. All numbers expressed in t CO_2 equivalents/year.

increased following the upgrade due to displacement of nitrogen and phosphorous fertilizers and reduction in N_2O from replacement of inorganic fertilizers (Willis *et al.*, 2018).

A year later, the team at Blue Plains took measurements of methane release during downstream storage and dewatering of the thermally hydrolyzed digested cake (Willis *et al.*, 2019). The authors proposed correlations to determine methane emissions from both points. The use of the correlations implies that while dewatering emissions were approximately five times higher than those in post-digestion storage, when compared with the carbon footprint reduction from using the biogas to generate renewable energy both were largely insignificant, and, based on typical density of methane, equivalent to leaks of 3.5–4% of the biogas produced, or a loss of around 7 kg CH_4/t raw sludge DSs digested per day (RDSD/d). These findings are similar to previously published work on advanced digestion with values between 5.6 and 6.2 kg CH_4/RDSD/d (Barber, 2007) which are lower than those for standard digestion (Barber 2007; UKWIR, 2005), which showed values between 10 and 21.5 kg CH_4/RDSD/d.

REFERENCES

Alexander R. and Peot C. (2017). Developing a Large-scale CAMBI Biosolids Marketing and Distribution Program. Proceedings of WEF Residuals and Biosolids Conference.

Asian Development Bank. (2012). Promoting Beneficial Sewage Sludge Utilization in the People's Republic of China, Asian development bank place, Beijing. ISBN 978-92-9092-654-2.

Auty D. and Blake J. (2009). Strategic Carbon Technology – How to Use Sludge to Your Advantage. 13th European Biosolids & Organic Resources Conference & Workshop.

Barber W. P. F. (2007). Use of an Elemental Balance Over Primary and Secondary Anaerobic Digestion of Sewage Sludge to Determine its Carbon Impact. Proceedings: 12th European Biosolids and Organic Resources Conference.

Barber W. P. F. (2009a) Influence of anaerobic digestion on the carbon footprint of various sewage sludge treatment options. *Water and Environment Journal*, **23**(3), 170–179.

Barber W. P. F (2009b). Calculation of the Carbon Footprint of United Utilities' Sludge Balanced Asset Programme. Proceedings of: 13th European Biosolids & Organic Resources Conference & Workshop.

Blytt L. D. (2009). The Cambi Process and Agriculture, A Sustainable Solution? Proceedings of Aquaenviro 14th European Biosolids and Organic Resources Conference and Exhibition, Leeds, UK.

Brandt R. C. (2018). Odor Characterization of Biosolids Products. Presented at Workshop B: High Quality Biosolids: Using Odor Predictors and Maintaining Resilient Biosolids Products to Expand and Sustain Customers, WEF Residuals and Biosolids Conference.

Brower B., Peot C. and Razik A. (2017). A commitment to improving the odors of DC water biosolids leads to improved acceptance. *Proceedings of the Water Environment Federation*, **2017**(1), 234-246.

Brower B., Peot C. and Romero A. (2018). Developing products with thermally-hydrolyzed biosolids at DC water. *Proceedings of the Water Environment Federation*, 2018(4), 1-14.

Elliott H. A., Taylor M., Brandt R. C., Hile M., Parsons J. J. and Niu X. M. (2018). Biosolids Odor: Processing Effects and Predictive Modelling. Presented at Workshop B: High Quality Biosolids: Using Odor Predictors and Maintaining Resilient Biosolids Products to Expand and Sustain Customers, WEF Residuals and Biosolids Conference.

Hammond G. and Jones C. (2008). "Inventory of Carbon and Energy (ICE) – Version 1.6". Sustainable Energy Research Team, University of Bath, UK.

Higgins M. J., Adams G., Chen Y. C., Erdal Z., Forbes Jr R. H., Glindemann D., Hargreaves J. R., McEwen D., Murthy S. N., Novak J. T. and Witherspoon J. (2008). Role of protein, amino acids, and enzyme activity on odor production from anaerobically digested and dewatered biosolids. *Water Environment Research*, **80**(2), 127–135.

Inglezakis V. J., Zorpas A. A., Karagiannidis A., Samaras P., Voukkali I. and Sklari S. (2014). European Union Legislation on sewage sludge management. *Fresenius Environmental Bulletin*, **23**, 635–639.

Lancaster R. (2015). Thermal Hydrolysis at Davyhulme WWTW, One Year On. Water Environment Federation (WEF) Residuals and Biosolids.

Lester J. N. (1987a). "Heavy Metals in Wastewater and Sludge Treatment Processes: Volume I, Sources, Analysis, and Legislation". CRC Press, Boca Raton, Florida.

Lester J. N. (1987b). "Heavy Metals in Wastewater and Sludge Treatment Processes: Volume II, Treatment and Disposal". CRC Press, Boca Raton, Florida.

Lester J. N. (1987c). Biological treatment. In: Heavy Metals in Wastewater and Sludge Treatment Processes Volume II: Treatment and Disposal, J. N. Lester (ed.). CRC Press, Boca Raton, Florida, pp. 15–40.

Murthy S., Higgins M., Chen Y. C., Novak J., Wilson C., Riffat R. and Aynur S. (2009). Impacts of Enhanced Digestion Processes on Biosolids Quality Parameters: Odors and Indicators. *Proceedings of the Water Environment Federation*, **2009**(12), 3936–3943.

Spalding S. and Smoot S. (2019). Worthless Dust or Valuable Resource? Drying Thermally Hydrolyzed Solids the Right Way. Proceedings WEFTEC Conference, Chicago, IL, pp. 4622–4633.

UKWIR. (2005). Workbook for Quantifying Greenhouse Gas Emissions (05/CL/01/3) UKWIR – UK Water Industry Research Centre.

Wang J-W., Li W., Jiang Q-H., Han X-Y, Hao S. and Tang C-C. (2018). Digestion enhancement with Thermal Hydrolysis Pretreatment at Beijing's Gaobeidian Water Reclamation Plant – Operating Experience and Challenges. *Proceedings of WEFTEC*, **2018**, 2821–2834.

Willis J., Bailey W., Jolly E., Murthy S., Peot C. and Trueblood D. (2018). DCWASA'S Certifiable GHG Inventory and Projected GHG Reductions from Cambi/Anaerobic Digestion Upgrades. *Proceedings of the Water Environment Federation*, **2010**(2), 141–147.

Willis J., Brower B., Domkreo H., Fotouhi J., Peot C. and Regmi P. (2019). New GHG Methodology for Estimation of Methane Emissions from Digested-Biosolids Storage and Dewatering. Proceedings of Water Environment Federation WEFTEC conference, pp. 5004–5011.

Chapter 6

Case studies

Do not read about this or that process, but go and look. I know instances where bitter enemies of chemical processes have been convinced of their error by just one unexpected and unprepared for visit of inspection

J. W. Slatter (1888)

The words above spoken by Slatter, written over 130 years ago about process technology in the sludge industry, remain as relevant today as they did in the Victorian era. This chapter highlights various case studies to give examples of how thermal hydrolysis has been used to influence the existing and new sludge treatment works, and expands on the benefits described in Chapter 5. However, while they provide an introduction, these are no substitute for visits to actual facilities to gain insight from the existing owners and discuss things that go well or not so well.

6.1 CASE STUDIES

A variety of case-studies have been chosen based on differing drivers and availability of information. A summary of the case-studies is given in Table 6.1. Although drivers may be different, it will become apparent that outcomes of the projects are similar.

Author: William Barber
doi: 10.2166/9781789060287_0163

Table 6.1 Summary of case studies.

Case-Study	Supplier	Size (t DS/d)	Configuration	Commissioning Date	Main Drivers
Davyhulme, UK	Cambi	330	Upstream 4 streams of 5 × B12 reactors	2012	Increase capacity in existing digestion and incineration plants High quality biosolids Reduced operating costs Reduced carbon footprint
Blue Plains, USA	Cambi	405	Upstream 4 streams of 6 × B12 reactors	2014	Reduced cost of proposed digestion plant Renewable energy production High-quality biosolids product Reduced operating costs Reduce risks associated with biosolids production
Billund, Denmark	Krüger-Veolia	15	Intermediate Exelys plug-flow	2015	Enhanced energy efficiency Co-digestion Show case innovation
Beijing	Cambi	>1,000	Upstream	2017–2019	Increase renewable energy production High quality biosolids product Increase capacity of existing digestion Reduce environmental impact

6.2 DAVYHULME, MANCHESTER, ENGLAND. OWNER: UNITED UTILITIES

6.2.1 Overview

United Utilities, a major UK Water Company, employed thermal hydrolysis at its largest digestion plant in Davyhulme to increase capacity at an off-site incineration plant connected via a sludge pipeline. In addition, installation of a dewatering facility and co-generation upgrades at the facility enables land recycling of biosolids cake while producing renewable energy. Through its link with the incineration plant, sludge processed through Davyhulme can be recycled as enhanced treated biosolids from there or sent through the pipeline to the incineration plant where it can be dewatered and recycled as a standard quality biosolids product or burnt in the incinerator. Subsequently, the plant is the UK's largest producer of renewable energy from sewage-derived biogas and has won awards for sustainability. Recent work has seen the biogas produced being upgraded to biomethane which can be injected onto a gas grid or used in vehicles.

6.2.2 Background

In the late 1990s, restrictions on nitrogen application; concerns over metals; changes to farming practices; public perception and reduction in brown-field reclamation were major factors influencing the use of biosolids to agriculture in the UK. Independent studies commissioned by United Utilities confirmed the potential landbank reduction for its catchment area covering the North West of England. Consequently, in 2002, United Utilities formulated the first of several strategies to reduce its reliance on land application from around 70% to circa 40% by the installation of additional incineration. This was to be achieved by upgrading capacity at its existing digested biosolids incineration facility and by constructing a second incineration plant for raw sludge, which up to that point, was being lime treated. However, major concerns within the team tasked with delivering the new incinerator, along with drivers turning toward land application combined with increased environmental awareness raised doubt into the long-term sustainability of the proposed incinerator. Subsequently, United Utilities undertook further studies looking at many potential solutions for the project and ultimately investigated two options along with the existing solution (of lime addition) in further detail. The eventual solution to the proposed project had to be one which (Lancaster, 2015):

- Maximized use of existing assets
- Significantly increased renewable energy generation
- Reduced the need for incineration
- Reduced United Utilities' operational carbon footprint and environmental impact

- Produced a consistently high quality biosolids product suitable for recycling in the North West of England
- Reduced customers' bills/rates
- Provided full contingency during maintenance periods and
- Aligned fully with long-term sludge strategy

Key to maximizing assets was necessary to increase the capacity of the existing incineration plant. This plant was positioned strategically on a 60-mile-long sludge pipeline which was fed liquid sludge from several digestion facilities. The pipeline was a legacy connected to a previous strategy whereby liquid digested sludge was directed toward the coast and dumped at sea. When this was outlawed in 1998, the incineration plant was installed, and sludge flow was redirected to it. As Davyhulme contributed approximately two-thirds of the sludge into the incinerator, Davyhulme was highly influential to its operation. Thermal hydrolysis was chosen at Davyhulme because it was the only technological option which could both double the loading rate in the existing digestion plant at Davyhulme, and increase capacity in the existing incineration plant. Doubling the loading rate enabled the sludge destined for the new incinerator to be redirected to Davyhulme where it could be processed, thereby removing the need for the new incinerator. By improving digestion performance, more biogas is produced therefore less biosolids is available for incineration. However, in order to increase the incinerator capacity, it was necessary to increase the energy content of the sludge. Tests were conducted to determine the influence of thermal hydrolysis on mixtures of the digested sludges on the pipeline, of which only the sludge provided by Davyhulme was thermally hydrolyzed. The tests showed that the dry solids (DSs) achievable was approximately the weighted average of the individual dewatering performances. By enhancing the dewatering of the Davyhulme sludge by the use of thermal hydrolysis, the overall DSs of the mixed digested sludges increased at the incineration plant. As described earlier, this results in more energy within the sludge cake. Improved dewatering resulted in less water and subsequently higher energy content on a cake basis. Sensitivity analysis showed that a dewatering improvement of 3–4 percentage points was required to offset the energy lost in the cake due to reduced volatile solids content brought about by improved digestion performance. Thermal hydrolysis was found to improve the dewatering by a greater quantity therefore increasing the energy content in the cake and ultimately incineration capacity.

Figure 6.1 shows the combustion diagram for the incinerator (Barber, 2009a). It is possible to burn any combination of throughput and energy within the shape which is limited by the geometry of the existing plant and defines the boundary operating conditions. The dashed line refers to the point where the material is autothermic and therefore requires no auxiliary fuel. Any operating point

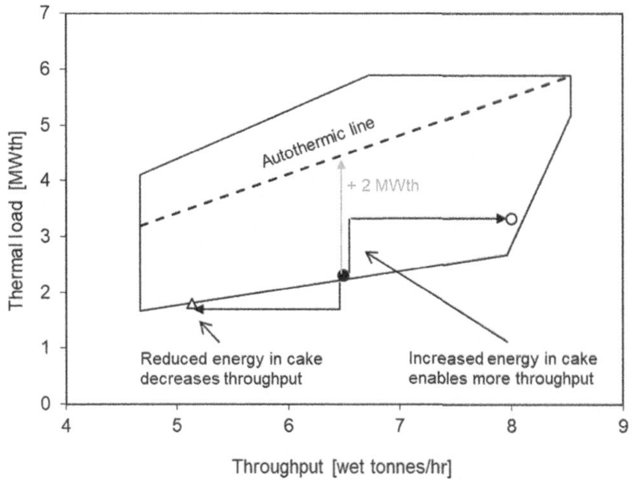

Figure 6.1 Incineration line diagram for streams 1 and 2, incinerator. *Key*: Black filled circle = typical operating point, 6.5 t/hr biosolids cake with energy content of 2.3 MW thermal energy; white filled circle = influence of thermal hydrolysis and improved digestion; white filled triangle = influence of advanced digestion (adapted from Barber, 2009).

below the line, therefore, requires additional fuel to reach the dashed line. In Figure 6.1, the black filled circle refers to a typical operating point, 6.5 t wet cake/hr with an energy content of approximately 2.3 MW thermal load. As shown, a further 2 MW energy are required to enable the biosolids to burn. If energy is put into the biosolids at an equivalent throughput, the operating point moves toward the direction of the autothermic line. At this point, throughput is not limited by the geometry of the incinerator and it is possible to shift the point to the right until it meets the edge of the combustion diagram. This is shown by the white filled circle in Figure 6.1 and shows the influence of thermal hydrolysis on digestion at Davyhulme. However, if dewaterability is not improved but upstream digestion is, then the energy content of the cake reduces and the operating point falls outside the confines of the incinerator. Now, it is necessary to modify the position of the operating point toward the left, until it rejoins the incinerator limitations. In this instance, capacity is lost. This analysis was fundamental in achieving increased incineration capacity for this project.

The project increased the incineration capacity by 14,000 dry t digested sludge annually of which 9,000 was due to improved dewatering and the remainder due to conversion to biogas upstream (Barber, 2009a, b). Figure 6.2 shows the development of United Utilities' sludge management plan over time.

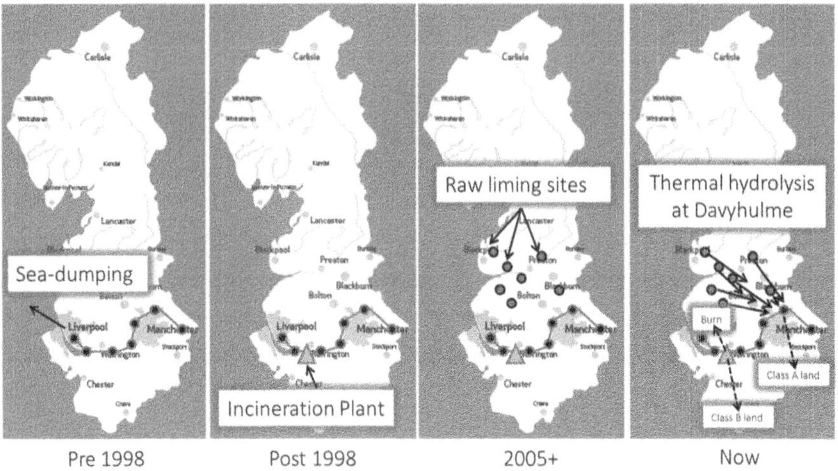

Figure 6.2 Development of United Utilities' sludge strategy. Prior to 1998 digested liquid sludge from seven sites was fed into a pipeline and dumped at sea. When this was outlawed it was burnt in a purpose-built facility based on dewatering and incineration. Subsequent project after 2005 looked at developing new incinerator for seven sites where liming was employed. That project evolved into one involving thermal hydrolysis to digest all sludge proposed for burning at an existing digestion plant at Davyhulme from where it could be recycled to land as Class A material, or sent into pipeline to Incineration plant where it could be dewatered and recycled as Class B material or burnt.

6.2.3 Project information and outcome

> *Installing thermal hydrolysis at Davyhulme significantly increased capacity of anaerobic digestion and incineration without building additional capacity for either*

A contract for 75 million GBP was awarded to Black and Veatch in 2010. The programme of works was based on sludge cake deliveries being made to Davyhulme, where the sludge is thermally hydrolyzed and digested alongside the Davyhulme sludge, then dewatered in a new facility for enhanced treated sludge recycling, or pumped as a liquid to the incineration plant where the sludge is dewatered or burnt. The thermal hydrolysis plant is shown in Figure 6.3, while the wider infrastructure in Figure 6.4. The dewatered cake import reception facility is based on 2 × 85 m^3 cake reception hoppers taking deliveries from 23 t trucks. Cake is transferred by elevated Chainlink conveyors (used also at the incinerator) to 2 × 800 m^3 gallon glass-lined storage vessels. The thickened sludge and cake is then blended within a range of 16–19% DSs and fed to a Cambi thermal hydrolysis plant made up of four trains of 5 × B12 reactors. The installed throughput of the thermal hydrolysis plant is approximately 330 dry t/d,

Figure 6.3 Thermal hydrolysis installation at Davyhulme, UK (courtesy: Cambi).

Figure 6.4 Thermal hydrolysis plant (in shaded area) at Davyhulme, relative to other major plant components (courtesy: Cambi).

of which circa 250 is design flow. Treated sludge is transferred to the existing digesters which have had additional spray polyurethane insulation to maintain the temperature differential between the inside and outside surfaces of the structures due to the higher operating temperatures of the sludge after it had passed through the THP plant (McNeill & Thornton, 2011).

Two 8,900 m^3 flexible membrane gas holders store biogas prior to treatment for siloxane and other contaminants before burning in the co-generation plant comprising 5 × 2.4 MWe Jenbacher CHP engines (McNeill & Thornton, 2011). Three of the engines existed and were refurbished and moved to accompany two new engines in a co-generation building.

The project has enabled approximately 45 t DS/d spare capacity in the incinerator (15 and 30 t DS/d due to improved digestion and dewatering respectively) and 165 t DS/d spare capacity in the Davyhulme digestion system, without building any further incineration nor digestion capacity.

The detailed design, construction and installation took 3.5 years, followed by 6 months commissioning and 1 year optimization (Edgington *et al.*, 2014). The commissioning of the facility started in late October 2012 with the liquid sludge import and screening plant. This was followed by biogas storage and cleaning plant with new engines. The three existing engines were moved between April and September 2013. The thermal hydrolysis plant and digestion facility were up and running within a 4-month period (Belshaw *et al.*, 2013). Experiences of which are provided in Chapter 4.

The plant has exceeded all guarantee parameters, notably a gas production of over 400 m^3/t DS fed, biosolids cake in excess of 31% DSs. Table 6.2 shows the

Table 6.2 Performance data* from thermal hydrolysis and associated equipment at Davyhulme.

Parameter	Units	Expected Performance	Actual Performance
Thickening polymer usage	kg/t DS	<10	6
Dewatering polymer usage	kg/t DS	<10	8
Throughput	t DS/d	250	203
Specific power consumption	kWhr/t DS	197	152
Renewable energy generation	kWhr/t DS	800	800–920
Cake DSs	%DS	>28.5	>31
Sludge production status	Pass/Fail	Enhanced	Enhanced
Specific biogas production	N m^3/t DS	400	400
Biogas utilization to boilers	%	8	4.1
Volatile destruction	%	60	58–64
Gas sent to co-generation	%	91	94

Taken from Lancaster (2015).

contractual guarantees for the project. The latter was due to the development of specific operations and knowhow developed by United Utilities (Lancaster, 2015). The award-winning facility (awarded 'Most innovated green energy scheme in the world' by the Institution of Chemical Engineering, 2013, and Major Civil Engineering Construction Project 2014, by the British Construction Industry) is the largest thermal hydrolysis plant in Europe, has reduced the utilities' carbon footprint by 8% (Barber 2009a, b), is the UK's greatest generator of renewable energy from sludge and has resulted in a significant reduction in the need for chemicals and makes operational savings of between 1.5 and 5 million GBP annually (Lancaster, 2015). Subsequently a biogas upgrading facility to pipeline quality has been installed (Lissett *et al.*, 2016), and United Utilities has installed two further Cambi thermal hydrolysis plants at Leigh and Burnley.

6.3 BLUE PLAINS, WASHINGTON DC, UNITED STATES OF AMERICA. OWNER: DC WATER

6.3.1 Overview

DC Water has installed the largest (at the time of writing) and first thermal hydrolysis plant in North America, at its plant in Blue Plains. The plant is combined with new anaerobic digestion and belt dewatering plants. Previous schemes to install anaerobic digestion were abandoned due to escalating costs, and detailed research programmes guided DC Water toward the use of thermal hydrolysis. Besides reducing the capital costs of the scheme by lowering the digestion capacity required, the plant has resulted in operational savings, a reduction in carbon footprint of 30% and the creation of a valuable biosolids product which is being marketed and sold under the registered trademark Bloom®. The facility has subsequently won various awards for sustainability.

6.3.2 Background

DC Water's Blue Plains treatment works, located 7 miles south of Washington DC in the United States, is designed to treat 1.4 million m^3 flow daily. As it's effluent contributes to the Chesapeake Bay, the largest estuary in the United States, the plant is required to meet discharge limits of 3 mg/l total nitrogen and 0.18 mg/l phosphorous. The plant has primary treatment followed by high-rate activated sludge system for carbon removal, prior to a tertiary treatment plant based on combined nitrification denitrification. Ferric chloride is used to meet the discharge standard for phosphorous.

With respect to sludge, a plan was developed in the late 1990s with three primary goals: first, to improve solids processing system reliability; second, to provide adequate treatment capacity for current and future needs and finally, to address community concerns, such as the odors. In September 1999, the Board adopted the recommendations of a Biosolids Master Plan (BMP) to anaerobically digest all the solids produced at Blue Plains. In the event of land application becoming unsustainable, a drying plant would be installed to provide alternative outlets. Meanwhile, in the year 2000, the existing digestion plant, which had struggled with capacity since the 1980s, was finally shut down after 70 years of operation since 1938. Following decommissioning of the digesters, biosolids processing was based on adding lime to dewatered cake prior to recycling on land as a Class B (standard treated equivalent to 2-log pathogen kill) material. I Approximately 1,100 wet t (65 trucks) of biosolids was recycled to agricultural outlets, as well as 16,000 ha of silviculture, poplar plantation and various land reclamation projects. Although farmers valued the material at $300/acre, the biosolids were recycled using third parties for a charge of approximately $43/wet t. Based on the tonnage, this amounted to $19 million/year of costs and comprised 21% of DC Water's annual operating budget for the works. There were a number of odor complaints and staff had worked on various initiatives to minimize odors over several years.

Soon after the digestion plant was closed, a BMP concluded that a new digestion plant should be installed based on using 8 × 17,000 m^3 egg-shaped anaerobic digesters (after discounting two-stage digestion), considered to perform better than other shaped digesters. Design work on the egg-shaped digester facility was conducted between 2003 and 2006. However, rapid escalation in steel prices, limited competition and other factors resulted in a project that was considered too costly and risky for the likely benefits that it would provide.

Subsequently, in late 2006, the Board of Directors deferred the egg-shaped digestion proposal while it conducted a review of the market place. By 2008, the previous plan was revised to provide a solution which was affordable, reduced current operating costs, energy-efficient, proven at large scale, environmentally sound and capable of producing Class A biosolids.

At this time, DC Water conducted research on numerous configurations of anaerobic digestion and these were benchmarked on variables such as: performance, financial impacts, greenhouse emissions, energy balances and odor potential. The work concluded that a solution based on thermal hydrolysis would meet the requirements of the project.

The key benefits of thermal hydrolysis were a reduction in digestion capacity required and critically cost of the digestion component of the project, production of renewable energy through biogas, reduction in biosolids production by approximately 50% due to a combination of improved dewatering and the avoidance of lime, and a higher quality of material produced.

6.3.3 Project information and outcome

In 2011, construction began on the project based on provision of: sludge blending tanks, sludge screens (by Hydro International), 10 × pre-dewatering centrifuges (Alfa Laval), cake bins and conveying, thermal hydrolysis plant comprising four

Figure 6.5 Photo of the thermal hydrolysis plant in DC Water (Photo by author).

streams of 6 × B12 reactors (Cambi – Figure 6.5), sludge cooling system, anaerobic digestion plant comprising 4 × 14,400 m^3 digesters with 15 d retention time at maximum throughput, new digested sludge dewatering building containing 16 dewatering belt presses with 2 m width belts (Simon Hartley Winklepress) and a co-generation plant based on 3 × 5 MW gas turbines (Solar).

The execution of the project was split into five distinct components, each with a different type of contract as follows: (1) site preparation (design, bid and build); (2) main process equipment (design build); (3) co-generation plant (design, build and operate – by third party); (4) odor control (design, bid and build) and (5) final dewatering plant (design, bid and build). The total project construction inclusive of engineering services was 470 million USD.

The plant started commissioning in late 2014, starting with the sludge screens. It took approximately 20 d to seed the digesters (from neighboring Alexandria, VA plant). Commissioning of other equipment took place later in 2014. It took 6 months to ramp up the digestion plant and stable performance was observed after 10 months of operation. Currently, the plant is achieving approximately 65% destruction of volatile solids, and the dewatering plant producing cake between 30% and 32% DSs. The associated polymer consumption is circa 10 kg/t. Biogas yields at the site range between 430 and 440 m^3/t DS fed to digester. This has resulted in the generation of 8 MW electricity which accounts for about a third of the plants total power consumption. Destruction of volatile solids, improved dewatering and the termination of lime use has reduced the annual production of biosolids by over half from 1,100 wet t/d to about 400 wet t/d. The carbon footprint of works at Blue Plains has reduced from 165,450 t CO_2 equivalents in

Figure 6.6 Fresh Bloom® foreground, and cured material in the background (courtesy: B. Brower).

2014 to 115,185 CO_2 equivalents after 2 years of operation at 2016 (Willis *et al.*, 2018), a site-wide reduction of 30%. The value of biogas combined with the reduction in cake biosolids recycling costs has amounted to operational savings of over 20 million USD.

DC Water is marketing the biosolids cake product as Bloom®. Fresh Bloom® is cake which has had no further processing and derived directly from the belt presses (Brower *et al.*, 2017); however, DC Water has discovered that the product properties can be further enhanced by increasing the DSs by either curing, or thermal drying, as shown in Figure 6.6, and described in Chapter 5.

DC Water has ambitious plans to sell a variety of biosolids products. The first product was generated from the digesters in late 2014, and 23 April 2018 was the first day that all biosolids produced were sold. Bagged material has been in stores from May 2018.

6.4 BILLUND BIOREFINERY, DENMARK. OWNER: BILLUND VAND A/S

6.4.1 Overview

Billund Biorefinery is a wastewater treatment plant which co-digests its sludge with various domestic and industrial organic waste streams. Due to the import of energy within the waste streams, the facility produces a surplus of energy. In addition, the facility undertakes clean water recovery from the wastewater it produces and recovers 98% of the waste materials that are made. Besides biogas and fertilizers, there is an intention to recover phosphorous and bioplastics. With respect to digestion, depending on the nature of the material being digested, it is either pasteurized and then anaerobically digested, or it is digested first followed by thermal hydrolysis and a second stage of digestion. In this way higher biogas yields are possible from the various waste materials. The thermal hydrolysis process is one which processes high DSs feed in a continuous system. The thermal hydrolysis unit is shown in Figure 6.7.

Figure 6.7 Thermal hydrolysis unit at Billund Biorefinery. (Courtesy: S. Viswanathan).

6.4.2 Background

The Billund BioRefinery is Public Private Partnership (PPP) between Krüger-Veolia, the municipality of Billund, and its utility company Billund Vand A/S, and is an extension of the wastewater treatment plant at Grinsdstead in Denmark. In 2013, a competition was organized and requested ideas on the 'Wastewater Treatment Plant of the Future' specifically to showcase Danish skills in wastewater treatment (Bro *et al.*, 2017). The PPP was set in co-operation with Danva (the Danish National Environmental Agency) and was supported by federal grants and 12 million USD.

The concept was based on optimizing the existing treatment works and introducing a number of technologies to further enhance plant performance. This was with the aim of improving water and nutrient reclamation while simultaneously reducing energy consumption and increasing the generation of energy. The plant treats 6,500 m^3/d of flow containing 0.65 and 0.1 t of nitrogen and phosphorous, respectively (Bro *et al.*, 2017).

The biorefinery had defined two major ambitions. The first one entailed the reclamation of water with minimal consumption of energy and consumables. This was achieved by improving aeration, installing membrane filtration, and the use of an intelligent online control system. In addition, adjustments were made to the sludge age in the activated sludge plant to divert more energy to the digestion system. The second ambition was related to the digestion of sludge and other waste materials (Figure 6.8). Table 6.3 shows the quantity and types of materials processed (adapted from Bro *et al.*, 2017). The total flow to the plant – comprising both site-generated sludge and imports – is approximately 50,000 wet t per year, equivalent to 5,800 dry t.

Sludge from the plant along with kitchen waste was treated by anaerobic digestion followed by thermal hydrolysis prior to a second stage of anaerobic digestion. The digestate from a first, thermophilic anaerobic stage, is dewatered to up to 25% DSs to minimize energy use (Thomson *et al.* 2017). However, there is no published data on the polymer required for this dewatering stage. Following this, the digestate is then thermally hydrolyzed. Unlike the Davyhulme and Blue Plains examples earlier, the thermal hydrolysis plant, Exelys – provided by Krüger-Veolia – is a continuous, rather than a batch process. The plant uses dynamic mixing in place of steam lances, and has a heat exchanger downstream

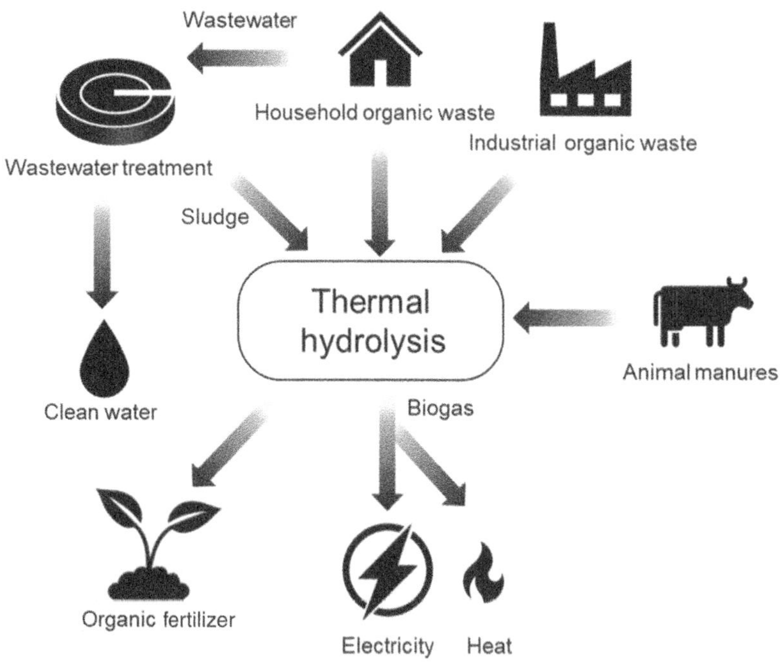

Figure 6.8 Concept behind Billund BioRefinery – adapted from Bro *et al.* (2017).

Table 6.3 Input waste streams to the Billund BioRefinery.

Material	Quantity (t DS/d)	DSs (%)	Volatile Solids (%)	COD Equivalence (kg/kg DS)	Treatment*
Primary sludge	2.1	5	70	0.70	D+TH+D
Activated sludge	1.4	5	67	0.50	D+TH+D
Kitchen waste	4.7	33	85	1.50	D+TH+D
Total	8.1	9.8**	71.6**	1.13**	
Flotation Fat	3.4	35	95	2.0	P+D
Fermentation residues	3.3	10	80	0.6	P+D
Other	1.2	10	75	1.0	P+D
Total	7.9	14.4**	81.5**	1.26**	
Overall total	16.0	11.6**	75.5**	1.2**	

*D+TH+D refers to anaerobic digestion followed by thermal hydrolysis and a second stage of anaerobic digestion. P+D refers to pasteurization followed by anaerobic digestion.
**Weighed average.

of the thermal hydrolysis plant which can pre-heat the material going in. It is purported that dynamic mixing provides a better steam distribution and helps lower steam demand than using steam lances as used on other variants of thermal hydrolysis. Sludge is held in the hydrolysis unit for 30 min, after which it is diluted and digested in a second mesophilic digester along with other materials which have been pretreated with pasteurization rather than thermal hydrolysis. The pasteurization process is based on heating the material at 70°C in batches for 1 hr to enable compliance with Danish Regulations for organic fertilizers (Bro *et al.*, 2017).

From Table 6.3, the typical input into the hydrolysis process is approximately 8 t DS/d (2,920 t DS annually), although the design capacity is 15.3 t DS/d, or 5,600 t DS per year (Bro *et al.*, 2017). In total, 350 m^3 of ammonia-rich digestate derived from the digestion process with average concentrations of 1,000–1,500 mg/l (equivalent to load of 350–525 kg/d ammonia) are processed in an AnitaMox® deammonification-based system. As with other deammonification systems, it is necessary to dilute the digestate such that ammonia levels are less than 1,500 mg/l.

6.4.3 Project information and outcome

The project was instigated in early 2013 and formally set up by June that year. Preliminary designs were conducted later between October and December. In April of the following year a pilot plant was formalized. Commissioning on the site began early in 2015 with commissioning of the new online control system and organics processing facility. Commissioning of the Exelys thermal hydrolysis plant started in August 2015 and the plant started up in early 2017.

Overall volatile solids destruction was 66% (Thomson *et al.*, 2017), and this resulted in a weighed average gas yield for all the wastes combined regardless of pre-treatment type of just over 500 m^3/t DS fed into the digestion system. The gas yield is boosted due to the addition of the imported materials as shown in Table 6.3. The biogas produced contains 7,936 GWhr of energy in the form of power and heat (Bro *et al.*, 2017) which is equivalent to 1,400 kWhr/t DS.

As the sludge can be thickened to 25%, the steam demand has dropped from 1 t/t DSs (Bro *et al.*, 2017) to 0.8 t/t DSs based on heat exchange between thermally hydrolyzed material and sludge entering the unit (Thomson *et al.*, 2017). A quantity of nearly 2,300 dry t of fertilizer is produced annually by the plant with average DSs of 24.3% (Thomson *et al.*, 2017). In 2014, the Billund BioRefinery won several awards for Water Reuse and sustainability.

6.5 BEIJING OWNER: BEIJING DRAINAGE GROUP COMPANY, LTD

6.5.1 Overview

Beijing Drainage Group Company, Ltd (BDG) responded to strong governmental reforms by implementing a strategy for biosolids management focused on enhancing production of renewable energy and generating a high value biosolids product for land recycling. Both result in a low environmental impact solution. Subsequently, BDG studied the concept of thermal hydrolysis with high DSs digestion from 2011 onwards. As of 2018, BDG owns five of the top 10 commercially operating thermal hydrolysis plants in the world, with a combined installed capacity of nearly 1,200 t DSs of sludge daily. Thermal hydrolysis is combined with plate press dewatering to overcome regulatory requirements enveloping the use of biosolids, and an in-house designed liquor treatment system based on deammonification. The new strategy has been highly successful, and the project at Huaifang – the largest underground advanced wastewater treatment plant in Asia with a design capacity of 600,000 m^3/d – won the IWA Gold Prize award for Exceptional Project Execution and Delivery.

6.5.2 Background

From initial concept, BGD have installed 5 of the top 10 largest thermal hydrolysis facilities in the world within only 7 years

Continuous economic growth in the China has led to rapid urbanization, environmental degradation and increased pressure on water resources. While China has made advances in construction of urban wastewater treatment for 600 major cities to address issues related to degradation of water courses and coastline, and this progress has highlighted concerns regarding water scarcity and dealing with ever increasing production of sewage sludge. Subsequently, the Chinese government has made strong commitment to reforms in the sector, to treat sludge and address water scarcity. There is also interest in increasing production of renewable energy from sources such as biogas. Nevertheless, anaerobic digestion of sludge has been difficult in China with low uptake. In 2015, municipal treatment works in China produced 70 million dry t of sludge annually but only 10% of those facilities had anaerobic digestion. Even then, only 70% were operational. High sand content, low volatile solids (typically 20–60% of DSs) and heavy metals have hampered the development of the technology, in spite of production of between 2 and 3 billion m^3 biogas.

Beijing Drainage Group Company, Ltd (BDG) provides treatment to over 30 million residents in Beijing. Before 2015, various treatment and disposal methods were used for sludge in Beijing including lime stabilization (75%), composting (11%), co-firing in cement kilns (6.5%) and thermal drying (7.5%) (Liao, 2016). Since then, BDG has responded to the Government's reforms by adopting a strategic-level management of its facilities and assets, based on the following:

(1) Enhancement of effluent quality for water reclamation, meeting regulatory requirements for surface water quality, producing approximately 1 billion m^3 water for reclamation annually
(2) Production of approximately 60 million m^3 biogas per year
(3) Production of greater than 550,000 t of biosolids for agricultural uses
(4) Use of heat pumps to provide heat for 10 million m^2 of building area
(5) Renewable energy production of over 60 million kWhr/year from a combination of solar panels and biogas
(6) Clean air in wastewater and biosolids management by odor control.

Following several studies, BDG followed a "1-5-1" policy to meet their strategic goals regarding sludge management (Zhang, 2016) as follows:

(1) *One* general technological solution for sludge treatment. Comprising thermal hydrolysis pre-treatment, high DSs anaerobic digestion, chamber filter press for dewatering and an in-house developed deammonification system for the treatment of reject liquors
(2) *Five* regional sludge centers located at:
    (a) Gaobeidian
    (b) Xiaohongmen
    (c) Huaifang
    (d) Qinghe II and
    (e) Gaoantun (Shi *et al.*, 2014)
(3) *One* beneficial outlet based on land application, i.e. seedling nursery, mine rehabilitation, sandy plot improvement, public gardening, forest soil improvement, top soil for landfill sites, etc.

Table 6.4 Summary of thermal hydrolysis facilities owned by Beijing Drainage Group Company, Ltd.

Site*	Size (t DS/d)	Configuration (no. of Streams × no. of Reactors)	Year of Commi-ssioning	Population Served	Digestion Capacity (m^3)
Gaobeidian	273	4 × 5	2017	4 million	62,800
Xiaohongmen	180	3 × 5	2017	3.5 million	60,000
Huaifang	180	4 × 4	2017	3.5 million	55,000
Qinghe II	160	3 × 4	2019	2 million	44,000
Gaoantun (with imports)	400	4 × 6	2017	4.5 million	88,000
Total	1,193	18 × 24		17.5 million	309,800

*All facilities by Cambi.

The design and construction of the five projects commenced from September 2014 and four of the five projects were completed by 2018. A summary of their configuration is given in Table 6.4.

6.5.3 Project information and outcome

The following project information is based on Gaobeidian as an example and is based largely on the informative study by Wang and co-workers (2018). Gaobeidian Wastewater Treatment Plant (WWTP) is the largest in Beijing. It has a treatment design capacity of 1,000,000 m^3/d. Plant effluent is reused or discharged to the Tonghui River (Wang *et al.*, 2018). The discharge standard requirements are 30, 15 and 0.3 mg/l for COD, total nitrogen and total phosphorus, respectively. Prior to 2008, Gaobeidian WWTP operated 16 anaerobic digesters installed in two phases. The original eight units were completed in 1993. They use methane biogas for mixing and operated under a mesophilic temperature range. A further eight digesters were subsequently constructed in 1999 and the plant was configured as a two-stage mesophilic digestion facility with 12 digesters for stage 1 and a further four digesters for stage 2. The effective volume of each digester is 7,850 m^3, with a hydraulic retention time of nearly 30 d. However, the original digesters were decommissioned due to construction issues. Prior to the project, a portion of the sludge was not anaerobically digested. Subsequently, in order to fully utilize Gaobeidian's existing anaerobic digestion plant and increase biogas production, BDG installed thermal hydrolysis as shown in Figure 6.9.

The thermal hydrolysis plant comprised of four streams of five reactors per stream but was also accompanied by: sludge pre-treatment inclusive of screening (to 3 mm) and storage; 12 × 70 m^3/hr centrifuge thickening; digester

Figure 6.9 Thermal hydrolysis installation in Gaobeidian, Beijing. Photo courtesy of Cambi.

gas infrastructure; 18 × plate press dewatering (as previously mentioned) and filtrate treatment system. For details on these units, the interested reader is referred to Wang *et al.* (2018). The anaerobic digestion plant was started up in September 2017, with the thermal hydrolysis plant commencing in February of the following year. By March of 2018, the entire sludge processing system was in full-operation. With thermal hydrolysis it was possible to treat the entire sludge flow in only 8 of the 16 digesters, and hydraulic retention time dropped from nearly 30 to approximately 20 d. Table 6.5, adapted from Wang's study, 2018, highlights the key performance criteria.

Importantly, for BDG, thermal hydrolysis assisted with meeting two key criteria. Firstly, it was necessary to meet 40% volatile solids destruction during digestion to pass regulatory requirements. While this was met prior to digestion, fluctuations in performance meant that there were times when this was difficult to achieve (Wang *et al.*, 2018). Secondly, there is a requirement to meet minimum DSs of 40% DSs for sludge and biosolids reuse and disposal. Several approaches have been used to meet this requirement including thermal drying, mixing in various materials and residues, liming and composting. As shown in Table 6.5, both parameters safely exceed the requirement after installation.

Expanding on previous work by AECOM (2012), Barber and co-workers presented a study looking at the influence of thermal hydrolysis on carbon footprint of Gaobeidian (Barber *et al.*, 2017). Various options were reviewed and compared

Table 6.5 Summary of performance data at Gaobeidian.

Parameter	Units	Before Installation	After Installation
Volatile solids destruction	%	Approx. 40%	53%
Biogas production	m^3/d	76,000	101,000
Biogas yield	$N\ m^3/t$ DS	280	372
COD destruction	%	50	70
Dewatering	%DS	18–20	>45
Fecal coliform	g/MPN		11
Ascaris eggs	% mortality		Below detection level

to the baseline (prior to thermal hydrolysis) some of which included thermal drying and composting to meet requirements of DSs *vide supra*. The model used was a revised version of the one first described by Barber (2009a, b), and determined carbon footprint by difference of contributory factors (such as power consumption greenhouse gas emissions, chemical consumption and transport) and beneficial factors (renewable energy generation, fertilizer displacement, carbon sequestration as examples). The Technical Options are described as follows:

(1) Thermal hydrolysis of only WAS, biosolids dewatered and thermally dried
(2) Primary sludge thickened to 7.5%, rest as (1) but different assumptions
(3) Thermal hydrolysis of both primary and WAS, biosolids dewatered and thermally dried
(4) As (1) but biosolids composted
(5) As (1) but no co-generation plant, biogas used for steam and then flared

Besides comparing to the baseline, the options were also compared to the eventual solution of thermal hydrolysis of all sludge followed by digestion, dewatering and land application of the biosolids. A summary of the findings is in Figure 6.10 *videlicet*.

As shown in Figure 6.10, options involving drying and composting increase the baseline carbon footprint by approximately 25%. The gas and electricity requirements for drying dominate overall carbon footprints where drying is part of the solution (over two-thirds of the footprint; Barber *et al.*, 2017). For the Technical Option with composting (Technical Route 4 in Figure 6.10) assumptions around transport were critical. Technical Route 5 (where biogas is not exploited for renewable energy) shows the influence of not using biogas. The solution chosen, reduced carbon footprint by over three quarters compared to the original solution. This was due to a combination of increased renewable energy; large reduction in biosolids quantity; production of a safe biosolids material with multiple outlets, and meeting the DSs requirements for reuse with no need for thermal drying.

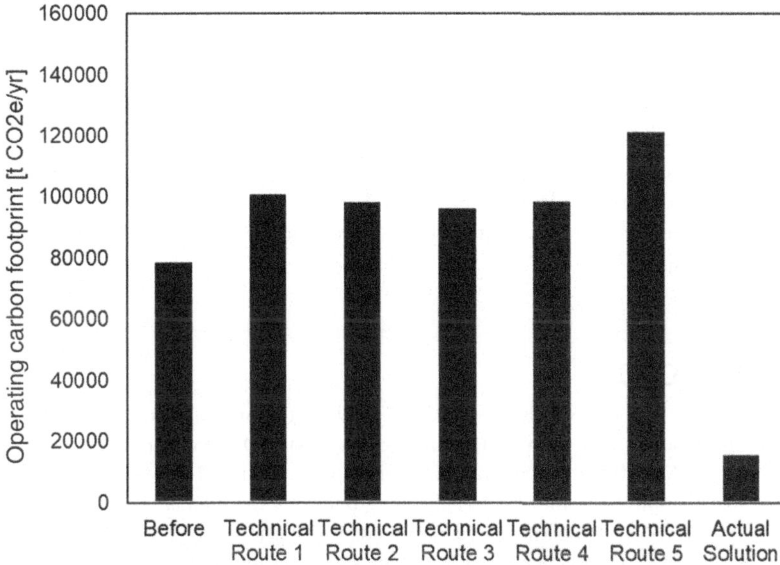

Figure 6.10 Influence of various biosolids solutions on carbon footprint of Gaobeidian both before and after implementation of thermal hydrolysis (actual solution). Adapted from Barber *et al.* (2017).

Subsequently, Beijing Drainage Group won an international award for Exceptional Project Execution and Delivery for its Huaifang Underground Water Reclamation Plant (HWRP) in 2018.

6.6 DISCUSSION

Although drivers may have been different, the facilities presented here – and the many which have not – share various commonalities. They have been led by owners looking at sludge management in a holistic manner and with longer term outlooks on future trends. As shown at the beginning of this chapter, thermal hydrolysis with anaerobic digestion is a good precursor to biosolids management regardless of outcome. By combining anaerobic digestion with technology which further improves it and improves dewaterability, more energy is extracted from the sludge and less biosolids are produced. This reduces further process requirements regardless of what they may be. The production of less biosolids combined with greater renewable energy production results in a process which has lower environmental impact and carbon footprint (as described in Chapter 5, Section 5.5). Furthermore, producing a sterilized low-odor material in lower quantities reduces long-term risks for biosolids management. By maximizing biogas production, systems can approach energy independency and reduce exposure to volatile energy markets and pricing. Risk is reduced further by the capacity release of existing digestion and other sludge infrastructure, allowing for future growth in sludge production. In addition, for clients with multiple facilities, after installation of thermal hydrolysis at one site, it habitually becomes a hub for other surrounding facilities even if not originally designed that way, thereby providing a level of flexibility with options at other sites. Clearly, the installation of thermal hydrolysis is site and driver-specific, and may not be a fit for all sites; however, it has continuously proven to be a valuable strategic asset to existing owners.

REFERENCES

AECOM (2012). Support to evaluation of sludge treatment technologies for Beijing Drainage Group. Report RSC-C13352 (PRC), financed by Asian Development Bank.

Barber W. P. F. (2009a). From raw incineration to advanced digestion – influence of sustainability on meeting a sludge strategy. Proceedings 12th IWA-Sludge Conference: Sustainable Management of Water and Wastewater Sludges, Harbin, China.

Barber W. P. (2009b). Influence of anaerobic digestion on the carbon footprint of various sewage sludge treatment options. *Water and Environment Journal*, **23**(3), 170–179.

Barber W. P. F., Liao Z., Kleiven H. and Fu X. (2017). Using Advanced Anaerobic Digestion to Minimize the Carbon Impact of Biosolids Processing: Case Study Gaobedian, Beijing. Proceedings of the 15th IWA World Conference on Anaerobic Digestion, Beijing, China.

Belshaw D., Edgington R. M. and Jolly M. (2013). Commissioning of United Utilities' thermal hydrolysis digestion plant at Davyhulme Waste Water Treatment Works. In Proceedings: 18th European Biosolids and Organic Resources Conference, Aqua Enviro Technology Transfer, UK.

Bro B., Raju C. S. and Gadegaard T. N. (2017). Billund biorefinery – how WWTPs can contribute to a circular economy. *Proceedings of the Water Environment Federation*, **2017**(15), 401–412.

Brower B., Peot C. and Razik A. (2017). A commitment to improving the odors of DC water biosolids lLeads to Improved Acceptance. *Proceedings of the Water Environment Federation*, **2017**(1), 234–246.

Edgington R. M., Belshaw D., Lancaster R. and Jolly M. (2014). Thermal Hydrolysis at Davyhulme WwTW – One Year on. Proceedings of: 19th European Biosolids & Organic Resources Conference & Exhibition, Aqua Enviro, UK.

Liao Z. L., Zhang R. B., Chen Y. J., Huang O. and Weng Y. Upgrading of Xiaohongmen WWTP to the first Cambi Advanced Anaerobic Digestion Project in Beijing – Design Construction, and Commissioning. Published and presented on European Biosolids and Organic Resources Conference, 15–16 November 2016, Edinburgh, Scotland.

Lissett T., Barnes L., Edgington R., Horne P. and Taylor J. (2016). Davyhulme Biomethane Utilisation Project maximising value for customers through integrated biogas energy generation. UK Water Projects 2016, pp 1–5.

McNeill J. and Thornton J. (2011). Davyhulme WwTW delivering United Utilities' sludge balanced asset programme. UK Water Projects, pp 105–108.

Shi L., Chen Y. J., Li Y., Huang O., Liao Z. L., Kleiven M., Chao S., Panter K., Kleiven H. and Barber B. (2014). Strategy Development of Sludge Treatment and Disposal in Beijing based on Thermal Hydrolysis Advanced Digestion. Proceedings of and presented at the 19th European Biosolids & Organic Resources Conference, 18–19 Nov, Manchester, UK, 2014.

Thomson C., Dimassimo R., Clay R., Viswanathan S., Kim J. and Landes N. (2017). Batch vs. continuous thermal hydrolysis – which is right for you? *Proceedings of the Water Environment Federation*, **2017**(1), 106–118.

Wang J.-W., Li W., Jiang Q.-H., Han X.-Y., Hao S. and Tang C.-C. (2018). Digestion enhancement with Thermal Hydrolysis Preatreamtn at Beijing's Gaobeidian Water Reclamation Plant – Operating Experience and Challenges. Proceedings of WEFTEC 2018, pp 2821–2834.

Westerholm M. (2013). Biogas production through the Syntophic Acetate-Oxidising Pathway. PhD thesis, Uppsala, Sweden.

Willis J., Brower B., Domkreo H., Gerbert L., Mulacek D., Peot C., Ramirez M. and Regmi P. (2018). Quantifying the GHG-emission reductions from DC water's digestion/Cambi/ CHP upgrades. *Proceedings of the Water Environment Federation*, **2018**(8), 4769–4775.

Zhang R. B., Sludge treatment and development in Beijing Drainage Group. Presented at The Third China Sludge Summit, Sept 2016, by China Biosolids Association, 9–11 Sept 2016, Zhengzhou, China.

Chapter 7

Economics

Insufficient attention is often given to the cost of operating and maintaining sewage treatment plants prior to their construction. Realistic appraisals of operating costs should precede the development of financing plans so that adequate funds are available...

P. P. Rowan, K. L. Jenkins and D. H. Howells (1961)

7.1 INTRODUCTION

Financial drivers play an important role in decision making with respect to biosolids processing. However, while important, decision making is multifaceted and involves numerous other influencing factors. Nevertheless, financial drivers comprise capital, operating, monetary (i.e. cost of borrowing) and other incentives, such as those in place to encourage the production of renewable energy. Concerning thermal hydrolysis, at a high level, three costs are the most relevant with respect to a project being financially viable. These being, the cost of manufacturing anaerobic digestion capacity, the cost of electricity/value of biogas, and finally the cost of recycling the biosolids. After these, other factors influencing costs include:

- Basic cost of capital equipment
- Impact on existing infrastructure. For example, can other equipment on site process more material at no/low extra cost?
- Requirement for future investment

Author: William Barber
doi: 10.2166/9781789060287_0187

- Requirement for contingency
- Requirement for additional liquor treatment
- Length of construction period
- Delays to project execution
- Time required to start up project for cost benefits to be realized
- Local versus foreign construction
- Local standards and requirements
- Use of unions
- Requirement and cost of consumables such as polymer, electricity, natural gas
- Cost volatility of consumables
- Incentives and subsidies for renewable energy and heat production
- Incentives and subsidies for biomethane production
- Competition for biosolids outlets
- Nutritional and other value attributed to biosolids which are land applied
- Gate fees for biosolids or food-waste, grease, or by-products from biosolids creation
- Long-term volatility of gate fees
- Cost of borrowing
- Insurance

These factors are geographically distinct making comparison between regions different. Furthermore, full-scale experience has shown that occasionally it may be financially beneficial to intentionally reduce or increase performance of various unit operations in order to see an overall cost saving. An example of this may be to intentionally reduce dewatering performance to save on polymer costs, as the cost increases in biosolids recycling may be lower than the cost saved on reducing polymer consumption. This is usually accommodated after the plant is constructed and operational, however, may be during a planning phase when choosing equipment and its sizing.

While thermal hydrolysis may be the critical component on a new biosolids management programme, as shown in Chapter 6, it may only contribute a small part of the additional infrastructure required to execute the project. Depending on the level of infrastructure available at an existing work where thermal hydrolysis is being considered, it may be necessary to additionally install:

- Preliminary treatment (screens)
- Sludge reception plant
- Sludge storage
- Conveying infrastructure
- Building and odor control (If necessary. Most thermal hydrolysis facilities are outdoors and use existing buildings for thickening, dewatering and other infrastructure.)
- Thickening plant
- Biogas infrastructure and treatment plant

- New or additional co-generation plant
- Biogas storage upgrades
- Biomethane upgrading if necessary
- Dewatering facility
- Liquor treatment plant
- Cake storage and conveyance

This list does not include other costs such as consulting fees, contractors, third parties, etc. Due to this, the actual cost of the thermal hydrolysis hardware (reactor vessels and associated ancillary equipment) may range from less than 5 (if digestion is also to be costed) to greater than 50% (for plants with no need to increase digestion capacity) of the overall project delivery cost. Thermal hydrolysis projects should also be viewed with respect to aborted future infrastructure projects. For example, DC Water's Blue Plains project saved money by not building four additional digesters. In contrast, installation of thermal hydrolysis at United Utilities' Davyhulme project resulted in the deferment of a new sludge incinerator.

7.1.1 Capital cost of thermal hydrolysis

The cost of a thermal hydrolysis unit can range from under 5% to over 50% of the whole cost of implementing a project depending on the level of infrastructure required

As would be expected, it is difficult to attain direct costs for thermal hydrolysis plants from suppliers. However, there are several third-party independent analyses which could be used for costing purposes. Unfortunately, if one were to plot the data from those studies, the data points would be scattered in a similar way to stars on a clear night sky. As previously mentioned, this is because thermal hydrolysis is a small—albeit key—part of a new biosolids infrastructure project. Most costing exercises do not present sufficient information on assumptions or provide detail on what components are used to make up the capital costs. In an informative UK study, Koodie and co-workers (2017) generated a series of cost curves for thermal hydrolysis and major associated infrastructure based on bottom-up estimating techniques. Costs were taken for projects ranging in size from approximately 15 to 90 t dry solids processed daily. The facilities all had existing anaerobic digestion and sludge-holding facilities, so these were not included in the costing exercise. Equations 7.1 and 7.2 (taken from Koodie *et al.*, 2017) show cost curves for thermal hydrolysis unit (equation 7.1) and thermal hydrolysis facility (equation 7.2). It should be noted that these equations are based on outturn costs inclusive of construction and installation. The curves were also based on costs from different suppliers of thermal hydrolysis.

$$C_{TH} = 0.3811\,Q + 9.5659 \quad (7.1)$$

where C_{TH} = cost of the thermal hydrolysis plant in GBP (in 2017), and Q = flowrate in tonnes dry solids/d. This equation was based on 14 cost quotes from different suppliers.

$$C_{THF} = 6 \times 10^6\, Q^{0.5509} \tag{7.2}$$

where C_{THF} = cost of the thermal hydrolysis facility in GBP (in 2017). This cost does not include the cost of anaerobic digestion plant or sludge-holding *vide supra.*

The work also highlighted the economies of scale by plotting unit costs (£ million/t DS) against plant size (Koodie *et al.*, 2017). Normalized facility costs dropped from approximately £2 million/t DS for plants <20 t DS processed/d to under £0.7 million/t DS at plant sizes over 100 t DS/d. The costs for the thermal hydrolysis unit itself also fell from £1 million/t DS processed daily to under £0.5 million/t DS over the same plant size range. It should be noted that cost analyses are geographically dependent and subsequently difficult to apply. Other studies have also shown costing data for thermal hydrolysis facilities (Abu-Orf & Goss, 2012; Childress *et al.*, 2019; Goss, 2015; Jolly & Gillard, 2009; Loomis *et al.*, 2016; Smith & Rus, 2019; Van Horn 2015; Williams, 2014) but some of these vary significantly even for projects of similar size and scope. The project in Medina, Ohio (Childress *et al.*, 2019) cost $35 million USD for a facility with a design capacity of approximately 20 t DS/d. In the absence of previous digestion, the facility included all infrastructure required for sludge treatment, inclusive of anaerobic digestion and recovery of phosphorous. By contrast, similar sized plants have cost more even though they were costed with less infrastructure (Abu-Orf & Goss, 2012), or without anaerobic digestion (Van Horn, 2015). Subsequently, there is a wide range in capital costs presented in the literature. From Goss (2015) Thames Water's plant at Longreach cost £23 million for a plant processing 36 t dry solids/d of activated sludge. This is equivalent to £1,750/t DS processed. Construction costs for Washington DC's facility rose above $400 million (Loomis *et al.*, 2016), equivalent to approximately $2,700 USD/t DS processed. When looking at capital costs, it becomes apparent that costs in the UK are lower than elsewhere.

Figure 7.1 expands on the work of Koodie & co-workers (2017) and shows an empirical relationship between normalized capital costs for thermal hydrolysis facilities and their size.

The graph can be described as follows:

$$C = 7.8066\, Q^{-0.509} \tag{7.3}$$

where C = cost of facility (£ million/t DS/d), Q = flowrate (t DS/d) ($R^2 = 0.7$).

The data points are based on plants with very different infrastructure requirements, and the use of equation 7.3 should be limited to high level economic evaluations. The equations presented here, are no substitute for site-specific budget costs.

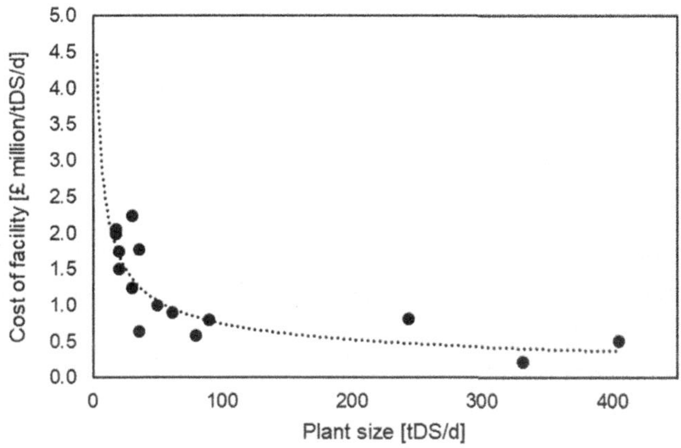

Figure 7.1 Capital costs for thermal hydrolysis facility inclusive of all ancillary plants. Note, the thermal hydrolysis cost itself will be between 5 and 50% of the cost in the chart.

7.1.2 Whole life cost example

This section describes a hypothetical example to give an indication of the operating costs of thermal hydrolysis. The example is then expanded upon to highlight various costing optimizations.

Question. *A treatment works processes 60 t DS/d of which 75% are volatile, and has an existing anaerobic digestion with three digesters of volume 6,400 m^3 each with a total volume of 19,200 m^3. Sludge is thickened to 5% using 3 kg polymer/t DS thickened. The digester processes mixed sludge and destroys 45% of the volatile solids. A total of 18,300 m^3/d of biogas is produced and used in a co-generation plant where low-grade heat is used to heat the digesters to 35°C. The plant generates 1.6 MW of electricity. The digested biosolids are dewatered to 22% dry solids using 8 kg polymer/t DS and recycled to land as a treated product. The utility owner needs to increase capacity to 100 t DS/d of the same type of sludge and wants to do this by installing thermal hydrolysis upfront of the existing digestion plant. What is the difference in the 30-year whole life cost between building new digestion capacity versus installing thermal hydrolysis? Use the following assumptions*:

- *Digestion performance with thermal hydrolysis = 62%*
- *Thermal hydrolysis of all sludge prior to digestion*
- *Thermal hydrolysis has 75°C temperature difference and is 65% efficient*
- *Gas yield = 0.9 m^3/kg VS destroyed*
- *Nitrogen content of volatile matter = 7%*
- *Maintenance cost is 1% of capital cost of thermal hydrolysis plant*

- *Thermal hydrolysis plant is 30% of overall plant cost*
- *For standard anaerobic digestion option 3 kg polymer/t DS and 9 kg polymer/t DS are required for thickening and dewatering, respectively*
- *For thermal hydrolysis 4 kg polymer/t DS and 10 kg polymer/t DS are required for thickening and dewatering, respectively*
- *Sludge thickened to 15% dry solids for thermal hydrolysis and to 5% dry solids for digestion. Unthickened sludge at 2% dry solids*
- *Post digestion dewatering performance is 32% dry solids for thermal hydrolysis with digestion and 22% dry solids for digestion. Capture rate is 95%*
- *Calorific value of biogas = 23,000 kJ/m^3*
- *There is an existing co-generation plant with electrical efficiency of 35%, high-grade heat 18%, low-grade heat 25%. Rest of the energy is lost. The availability of the plant is 90%*
- *Energy requirements of running co-generation plant, 5% of output electricity*
- *Assume natural gas is purchased to provide additional energy demand for thermal hydrolysis after high-grade heat is recovered from co-generation plant*
- *Assume waste heat recovered from co-generation plant is sufficient to heat digesters for standard digestion option (although this is unlikely to be the case with high proportions of biological sludge)*
- *Digester mixing requirement is 0.006 kW/m^3 (Metcalf and Eddy, 5th edition)*
- *Power requirements for thickening and dewatering (assuming centrifuges for both) = 1.1 and 0.9 kW/m^3/hr, respectively. Note: These are simplified data. Power requirements typically involve a fixed requirement combined with an additional requirement for throughput*
- *Power requirement of running thermal hydrolysis 15 kWhr/t DS*
- *Co-generation plant power requirements are 5% of the power generation*
- *Power requirements for biogas cleanup (siloxane, water, sulfide etc.) 0.26 kWhr/m^3 biogas*
- *Steam boiler efficiency 85%*
- *Cost of electricity is £0.08/kWhr*
- *Value of green energy is £0.14/kWhr*
- *Cost of natural gas is £30/MWhr (~£0.32/m^3)*
- *Cost of polymer is £2.50/kg*
- *Recycling cost of existing material is £40/wet t*
- *Recycling cost of enhanced treated material is £30/wet t*
- *Total nitrogen removal is required*
    - *Electricity required for N removal = 2 kWhr/kg N*
    - *Methanol required for N removal = 3.3 l/kg N*
- *Cost of building anaerobic digestion = £250/m^3*
- *Discount factor is 7%*

Answer.
Digestion capacity required

Based on loading rates given in Chapter 5, Table 5.1, assume loading rates are 2.5 kg VS/m^3 digester/d and 6.5 kg VS/m^3 for thermal hydrolysis pre-treated sludge.

For standard digestion

$$100 \text{ tDS/d} \times 75\% \text{ VS} \times 1{,}000 \text{ kg/t} = 75{,}000 \text{ kg/d are fed to the digester}$$

Therefore, volume required = 75,000/2.5 = 30,000 m^3
There is an existing volume of 19,200 m^3, therefore, a deficit of 10,800 m^3.
From the assumptions, this will cost £2,700,000

The mixing requirements for this plant will be 30,000 $m^3 \times 0.006 \text{ kW/m}^3$ = 180 kW

Per day, 180 × 24 = 4,320 kWhr

For thermal hydrolysis prior to digestion

75,000 kg/d are fed to the digester as before
Therefore, volume required = 75,000/6.5 = 11,538 m^3.
There is an existing volume of 19,000 m^3, therefore, no further anaerobic digestion capacity is required. In this instance it is possible to shut down one of the digesters, therefore, the used volume is 2 × 6,400 m^3 = 12,800 m^3.

The mixing requirements for this plant will be 12,800 $m^3 \times 0.006 \text{ kW/m}^3$ = 76.8 kW

Per day, 76.8 × 24 = 1,843 kWhr

Cost of thermal hydrolysis facility

Equation 7.3 is used to provide a high-level cost estimate as follows

$$C = 7.8066 \times 100^{-0.509} = \pounds 0.758 \text{ million/tDS/d} = \pounds 75{,}800{,}000$$

Cost of running thickening plant

(a) Polymer
For standard digestion 3 kg polymer/t DS are used, therefore, 300 kg/d. This costs £750/d.
For thermal hydrolysis with digestion, 4 kg polymer/t DS are used, therefore 400 kg/d. This costs £1,000/d.

(b) Power cost for thickening
For both cases, unthickened sludge is at 2% dry solids. Therefore, input volume is 5,000 m^3 (assuming density of water). Both instances with thickened sludge is 5,000 m^3/d or 208.3 m^3/hr.
Based on the assumptions, this will consume 229 kW = 5,496 kWhr/d.

Cost of running a thermal hydrolysis plant

(1) Cost for steam:

Steam consumption may be calculated from equation 2.2 (in Chapter 2) shown here

$$Q = (10^{(-0.464\,\mathrm{Ln}\,(D)+c)})/\eta$$

where Q = energy required (kJ) D = dry solids of sludge expressed as a percentage, and c is a constant which is a function of temperature difference ΔT and given by: $c = 3.8233\ (\Delta T)^{0.0786}$, and η = efficiency = 65%

$$Q = (10^{(-0.464\,\mathrm{Ln}\,(0.15)+c)})/0.65$$

$$c = 3.8233\ (\Delta T)^{0.0786} = 3.8233\ (75^{\circ}\mathrm{C})^{0.0786} = 5.368$$

substituting into the above:

$$Q = (10^{(-0.464\,\mathrm{Ln}\,(0.15)+5.368)})/0.65$$

$$Q = (10^{6.248})/0.65$$

$$Q = 2{,}724,\ 898\ \mathrm{kJ} = 757\ \mathrm{kWhr/t\,DS}$$

Therefore, for 100 t DS/d it is = 75,700 kWhr/d.

The quantity of natural gas required is simply the energy needed divided by the calorific value of methane.

= 75,700 kWhr/d/10.7 kWh/m^3 ('Fuel Gas.' *McGraw-Hill Encyclopedia of Science & Technology*. McGraw-Hill, Inc., 1982.)
= 7,075 m^3 natural gas/d

This would be the gas requirement if there was no heat recovered from the co-generation plant. In this instance, high-grade heat is available from the plant. The digester biogas contains 267,375 kWhr daily (*see later*). From the assumptions, the co-generation plant is available 90% of the time, and converts 18% of the energy in the biogas to high-grade heat.

Therefore, high-grade heat available for recovery

= 267,375 kWhr/d × 90% available
× 18% conversion to high-grade heat
= 43,314 kWhr/d.

Therefore, the actual energy required by thermal hydrolysis is

= 75,700 − 43,314
= 32,386 kWhr/d

The concomitant natural gas requirement is given by

$= 32{,}386\ \text{kWhr/d}/10.7\ \text{kWh/m}^3\,{}^* = 3{,}027\ \text{m}^3$natural gas/d

However, the boiler efficiency is 85%, therefore, the actual demand

$= 3{,}027\ \text{m}^3/85\%\ \text{efficiency} = 3{,}561\ \text{m}^3$

The cost of natural gas is £0.32/m^3. Therefore, daily gas cost is = £1,139.58.

(2) Electrical costs

Based on 15 kWhr/t DS, which is 15 kWhr/t DS × 100 t DS/d = 1,500 kWhr/d.

(3) Maintenance costs

The thermal hydrolysis plant is 30% of the overall facility cost, therefore, the total cost is £22,740,000.

1% of the thermal hydrolysis plant costs = £227,400/year

Digester performance and gas production

For standard digestion assuming performance is equivalent to before as sludge being fed is of the same composition

(1) Gas production

100 t DS/d enters the digestion plant with 75% volatile matter.

45% of the volatiles are destroyed, that is 100 t DS/d × 75% volatile × 45% destruction × 1,000 kg/t = 33,750 kg.

Based on the assumption given, the gas production = 33,750 kg VS × 0.9 m^3/kg VS = 30,375 m^3 biogas/d.

There is a yield of 303 m^3/biogas/t DS_{fed}, which compares with Table 5.2, Chapter 5.

The energy content of biogas = 30,375 m^3 biogas/d × 23,000 kJ/m^3 = 698,625,000 kJ/d = 194,063 kWhr.

The energy required to clean up gas

$= 0.26\ \text{kWhr/m}^3 \times 30{,}375\ \text{m}^3\ \text{biogas/d} = 7{,}897.5\ \text{kWhr}.$

The renewable energy produced is determined from the product of engine efficiency and availability:

$= 194{,}063\ \text{kWhr/d} \times 35\%\ \text{efficiency} \times 90\%\ \text{availability}$

$= 61{,}130\ \text{kWhr/d}\,(= 2.54\ \text{MW}_e)$

---

*Calorific value of natural gas in kWhr/m^3.

From assumptions, the energy required to run a co-generation plant is 5% of the output

$$= 61{,}130 \text{ kWhr/d} \times 5\%$$

$$= 3{,}057 \text{ kWhr/d}$$

(2) Sludge exiting digester

The mass of sludge exiting digester = 100 t DS/d − 33.75 tVS destroyed/d = 66.25 t/d.

Of this quantity, 41.25 t/d are volatile solids (i.e. 62% of total solids).

The dry solids exiting the digester are 3.31% based on a throughput of 2,000 m^3/d (=100 t DS/d/5% DS)

(3) Nitrogen released during digestion

From Chapter 4 the quantity of nitrogen entering the digester is

$$= 100 \text{ t dry solids/d)} \times 75\% \text{ (volatile fraction)} \times 7\% \text{ (nitrogen)}$$

$$= 5.25 \text{ t.}$$

If this is converted to ammonia and 20% remains within the biosolids, then the release of nitrogen is

$$= 5.25 \text{ t/d} \times 45\% \text{ (volatile solids destroyed)} \times (1 - 0.2)\text{(released)}$$

$$= 1.89 \text{ t.}$$

If the digester is fed at 5% DS, then this quantity gives a concentration of 945 mg/l.

For thermal hydrolysis with digestion

(4) Gas production

100 t DS/d enter the digestion plant with 75% volatile matter.

62% of the volatiles are destroyed, that is 100 t DS/d × 75% volatile × 62% destruction × 1,000 kg/t = 46,500 kg.

Based on the assumption given, gas production = 46,500 kg VS × 0.9 m^3/kg VS = 41,850 m^3 biogas/d.

This is a yield of 419 m^3/biogas/t DS_{fed}, which compares with Table 5.2, Chapter 5.

The energy content of biogas = 41,850 m^3 biogas/d × 23,000 kJ/m^3 = 962,550,000 kJ/d = 267,375 kWhr.

The energy required to clean up gas

$$= 0.26 \text{ kWhr/m}^3 \times 41{,}850 \text{ m}^3 \text{ biogas/d} = 10{,}881 \text{ kWhr.}$$

The renewable energy produced is determined from the product of engine efficiency and availability:

$$= 267{,}375 \text{ kWhr/d} \times 35\% \text{ efficiency} \times 90\% \text{ availability}$$

$$= 84{,}223 \text{ kWhr/d} (= 3.51\text{MW}_e)$$

From assumptions, energy required to run co-generation plant is 5% of the output

$$= 84{,}223 \text{ kWhr/d} \times 5\%$$
$$= 4{,}211 \text{ kWhr/d}$$

(5) Sludge exiting digester

The mass of the sludge exiting digester = 100 t DS/d − 46.5 tVS destroyed/d = 53.5 t/d.

Of this quantity, 28.5 t/d are volatile solids (i.e. 53% of total solids).

The dry solids exiting the digester are 5.35% based on a throughput of 1,000 m^3/d (=100 t DS/d/10% DS).

(6) Nitrogen released during digestion

The quantity of nitrogen entering the digester is 5.25 t, as for the base-case scenario.

If this is converted into ammonia and 20% remains within the biosolids, then the release of nitrogen is

$$= 5.25 \text{ t/d} \times 62\%(\text{volatile solids destroyed}) \times (1 - 0.2)(\text{released})$$
$$= 2.60 \text{ t (an increase of 37.5\%)}$$

If the digester is fed at 10% DS, then this quantity gives a concentration of 2,604 mg/l (an increase of 2.75 times).

Dewatering and cake production

(1) Dewatering power consumption

For the digestion option, volume exiting digestion is 2,000 m^3/d, or 83 m^3/hr.

From the assumptions, the power required by dewatering = 83 m^3/hr × 0.9 kW/m^3/hr = 74.7 kW = 1,793 kW hr/d.

For the option based on thermal hydrolysis with digestion, volume exiting digestion is 1,000 m^3/d, or 41.7 m^3/hr.

From the assumptions, the power required by dewatering = 41.7 m^3/hr × 0.9 kW/m^3/hr = 37.5 kW = 900 kWhr/d.

(2) Dewatering polymer consumption

For the digestion option, dry solids exiting digestion are 66.25 t DS/d.

From assumptions, daily polymer consumption = 66.25 t DS/d × 9 kg polymer/t DS = 596.25 kg/d.

Cost of polymer = £1,490.63/d

For the option based on thermal hydrolysis with digestion, dry solids exiting digestion are 53.5 t DS/d.

From assumptions, daily polymer consumption = 53.5 t DS/d × 10 kg polymer/t DS − 535 kg/d.

Cost of polymer = £1,337.50/d

(3) Biosolids cake production

For digestion option, 66.25 t DS/d will dewater to 22%DS with 95% capture rate.

Therefore, cake produced = (66.25 t DS/d × 95% capture)/(22% DS) = 286 wet t/d.

This is a ratio of 2.9 of wet cake compared to dry solids being processed, and lies within the typical range given in Chapter 5, Table 5.3.

From assumptions, this cake will cost £11,440/d to recycle.

For thermal hydrolysis option, 53.5 t DS/d will dewater to 32%DS with 95% capture rate.

Therefore, cake produced = (53.5 t DS/d × 95% capture)/(32%DS) = 158 wet t/d.

From assumptions, this cake will cost £4,740/d to recycle.

This is a ratio of 1.6 wet cake compared to dry solids being processed, and compares with thermal hydrolysis of all sludge prior to digestion shown in Chapter 5, Table 5.3. Table 7.1 summarizes the influence of the two options.

Based on the calculations above, the following charts show the breakdown of daily costs for both options. The annual operating costs are approximately £6.45 million and £5.06 million for digestion and digestion with thermal hydrolysis options, respectively. The breakdown of these costs are given in Figure 7.2.

Typically, in the absence of thermal hydrolysis, charges associated with cake recycling dominate the operating costs of treatment, as shown in Figure 7.2a). With thermal hydrolysis, cake recycling is still the major contributor, but polymer and electricity are the next most influential. These become more so in instances where total nitrogen is not required, as this will eliminate the carbon costs and a fraction of the electricity. This breakdown is consistent with earlier analyses (Batstone *et al.*, 2008; Fountain, 2019). Although a great deal of emphasis is given on the requirement of energy for thermal hydrolysis, the analysis shows, in agreement with full-scale plant operation, that the costs of providing this energy are not significant, largely due to the low cost of natural gas compared to carbon for denitrification or polymer for dewatering. An example will be shown later of the impact of increasing dry solids to reduce energy demand.

With respect to the operating costs, these are offset by value attributed to renewable energy from biogas. In this example, the value of renewable energy is £3.123 and £4.303 million for options without and with thermal hydrolysis, respectively. When these are considered, the operating costs are £3.329 million and £0.753 million for the two options. Therefore, in this example, thermal hydrolysis reduces the overall operating costs by almost £2.6 million annually. It should be remembered that this reduction in the annual operating cost is accompanied by the production of a higher quality of biosolids product (as described in Chapter 5).

Table 7.1 Difference in performance between digestion option and that involving thermal hydrolysis.

Item	Unit	Digestion Option	Thermal Hydrolysis Option
Electricity for thickening	kWhr/d	5,500	5,500
Polymer for thickening	kg/d	300	400
Anaerobic digestion requirement	m^3	30,000	11,538
Electricity for digester mixing	kWhr/d	4,320	1,843
Initial energy required for TH* steam	kWhr/d		75,704
Electricity required for TH ancillaries	kWhr/d		1,500
Biogas production	m^3/d	30,375	41,850
Biogas yield	m^3/t DS_{fed}	304	419
Energy in biogas prior to co-generation	kWhr/d	194,063	267,375
Energy required to clean biogas	kWHr/d	7897.5	10,881
Renewable energy production	MW_e	2.55	3.51
Outstanding energy required for TH steam after recovery of high-grade heat	kWhr/d		32,389
Natural gas required for TH after HGH recovered	m^3/d		3,561
Electricity required to run co-generation plant	kWhr/d	3,056	4,211
Electricity for dewatering	kWhr/d	1,800	900
Polymer for dewatering	kg/d	596.25	535
Nitrogen released requiring treatment	kg/d	1,890	2,604
Biosolids cake produced	wet t/d	286	159

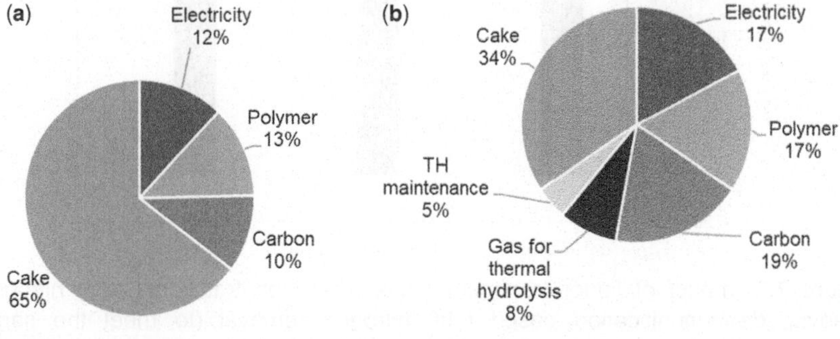

Figure 7.2 Breakdown of costs. (a) Continuation with digestion; (b) continuation with thermal hydrolysis.

Assuming no cost inflation, a 30-year NPV is approximately £90 million for the option with digestion, and £41 million when thermal hydrolysis is present. Over the lifetime of the project these numbers are equivalent to £81/t DS processed for the digestion option and £37/t DS processed for the option without digestion. As previously mentioned, the installation of thermal hydrolysis is not only about a financial decision, but one based on risk abatement and other factors as described in Chapter 5.

7.1.2.1 Influence of liquor treatment

> *Requirements and cost of liquor treatment are significant when looking at the economics of thermal hydrolysis*

This example is based on total nitrogen removal and is therefore skewed by the costs of carbon required. In many instances, such as in Europe, plants have a requirement to remove ammonia under the Urban Wastewater Treatment Directive (1991). Figure 7.3 shows the difference in annual operating cost if nitrification, and not total N is required, and this is also compared to deammonification for total nitrogen removal.

Figure 7.3 shows how influential liquor treatment requirements are on the overall cost of processing. As well as having a high cost, the use of methanol also significantly increases the overall carbon footprint of processing (Willis *et al.*, 2015). Note, the graph does not include for additional staffing requirements and does not account for differences in complexity between traditional treatment and deammonification which may have an influence on economics and operation.

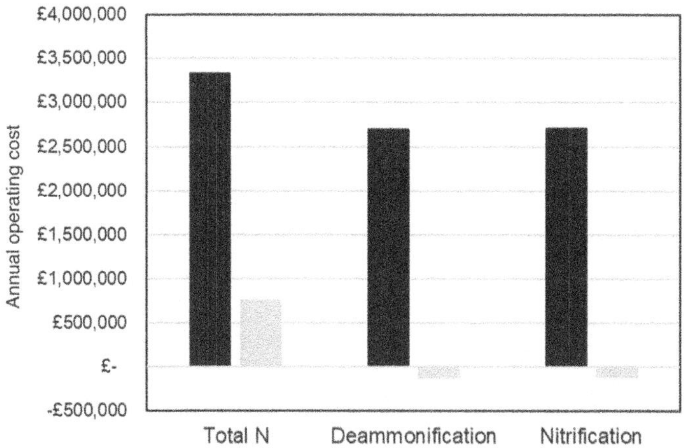

Figure 7.3 Impact of liquor treatment costs based on traditional total nitrogen removal, deammonification based total nitrogen removal (to meet the same requirements) and nitrification. Key: Anaerobic digestion (black bars); with thermal hydrolysis (gray bars).

7.1.2.2 The influence of polymer and steam consumption

The energy requirement for steam is routinely cited as a major influence in the running costs of thermal hydrolysis. As shown in the analysis, this influence is lower than expected. As shown in Chapter 2, it is possible to reduce the energy requirement for hydrolysis by increasing the dry solids entering the unit. Nonetheless, this comes at the expense of additional polymer. This is especially relevant if there is a large contribution from biological sludge in the mixture. In thickening and dewatering trials, Moncholi *et al.* (2018), described previously by Goss *et al.* (2017) demonstrated that 16–18% TS cake could be achieved with 9–13.5 kg active substance of dry polymer/t processed. They noted that dryer cake of 20–22% TS could be produced using emulsion polymer, but that required doses greater than 18 kg active substance/t DS. Figure 7.4, plot from various sources, gives an empirical relationship between polymer consumption and dewaterability for non-thermally hydrolyzed or pre-treated sludge.

The graph shows the following relationship ($R^2 = 0.943$)

$$P = 1922.2\,DS^{3.617} \tag{7.4}$$

where P = polymer consumption (kg active substance/t dry solids thickened; DS = dry solids required as a decimal (%).

Question. *Based on the above example, what would the change in operating cost be if the sludge was thickened to 25% dry solids but with a temperature difference of 85°C for thermal hydrolysis?*

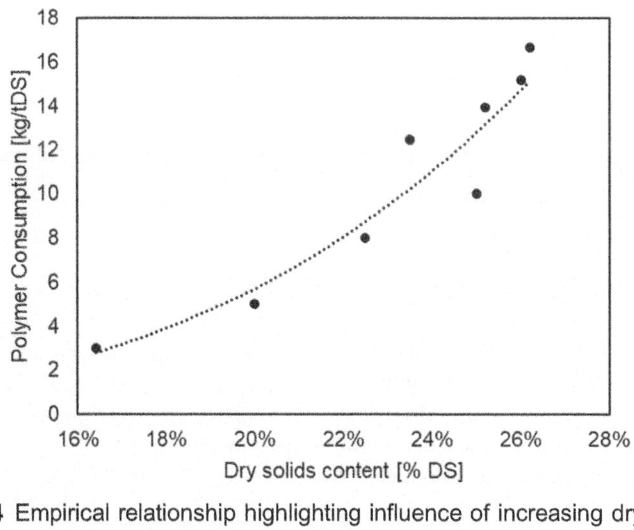

Figure 7.4 Empirical relationship highlighting influence of increasing dry solids for thickening on polymer consumption.

Answer.
From equation 7.4, at 15%, the polymer dose required = approximately 2 kg active substance/t DS, and to achieve 25% dry solids, 12.8 kg active substance are required.

From earlier, 3,561 m^3 natural gas/d are required after accounting for high-grade heat recovery.

Repeating the calculations for 25% dry solids to determine energy savings:

$$Q = (10^{-0.464\,\mathrm{Ln}\,(0.25)+5.421\dagger})/0.65 = 1{,}783{,}722\ \mathrm{kJ} = 495\ \mathrm{kWhr/t\,DS}$$

For 100 t DS/d it is 49,500 kWhr/d

Recovering waste heat of 43,314 kWhr leaves an outstanding energy demand of 6,186 kWhr/d.

This will require 680 m^3 natural gas/d after the boiler losses are accounted for.

As expected, the higher dry solids reduce the steam demand for thermal hydrolysis by approximately a third. The economic impact of doing this is shown, namely

At 15DS%

$$\text{gas cost} = 3{,}561\ \mathrm{m}^3 \times £0.32/\mathrm{m}^3 = £1{,}139.52/\mathrm{d}$$

and

$$\begin{aligned}\text{polymer cost} &= 2\ \text{kg polymer/t DS} \times 100\ \text{t DS/d} \times £2.50/\text{kg polymer}\\ &= £500/\text{d}\end{aligned}$$

Therefore,

$$\text{daily cost} = £1{,}139.52 + £500 = £1{,}639.52/\text{d}$$

At 25%

$$\text{gas cost} = 680\ \mathrm{m}^3 \times £0.32/\mathrm{m}^3 = £217.60/\mathrm{d},$$

and

$$\begin{aligned}\text{polymer cost} &= 12.8\ \text{kg polymer/t DS} \times 100\ \text{t DS/d} \times £2.50/\text{kg polymer}\\ &= £3{,}200/\text{d}\end{aligned}$$

Therefore,

$$\text{daily cost} = £217.60 + £3{,}200 = £3{,}417/\text{d}$$

In this example, increasing dry solids into the thermal hydrolysis unit has reduced the cost of steam energy by approximately £800/d, but this saving is offset by an increase in polymer cost

With thermal hydrolysis, polymer consumption, rather than heat requirement, has a larger influence on whole life cost

[†]New value, determined from equation 2.2, Chapter 2 to account for new temperature difference.

of approximately £2,500/d. This example highlights the importance of polymer use in pre-thermal hydrolysis dewatering. Therefore, when considering running thermal hydrolysis at higher dry solids, it is crucial that increased polymer requirements and subsequent costs are also accounted for.

7.1.2.3 Reducing performance of dewatering to lower costs

The following example looks at the potential for detuning dewatering performance.

Question. *The utility have been undertaking jar tests to optimize dewatering and find that they can reduce polymer consumption by 2 kg/t DS, but the dewatering performance will drop by 2 percentage points. Is it economic to do this?*

Answer.
In the original example there are 158 wet t of cake produced, and this requires 535 kg/d polymer. The daily costs of this are:

$$= 158 \text{ wet t/d} \times £30/\text{wet t} + 535 \text{ kg polymer} \times £2.50$$
$$= £4{,}740 \text{ (biosolids cake)} + £1{,}337.50 \text{ (polymer)} = £6{,}077.50$$

If dewatering is intentionally reduced to save the polymer.

The cake production increases to (53.5 t DS/d × 95% capture)/(30%DS) = 169 wet t/d

The polymer consumption reduces to 53.5 t DS/d × 8 kg polymer/t DS = 428 kg/d.

The new costs are

$$= 169 \text{ wet t/d} \times £30/\text{wet t} + 428 \text{ kg polymer} \times £2.50$$
$$= £5{,}070 \text{ (biosolids cake)} + £1{,}070 \text{ (polymer)} = £6{,}140$$

In this instance, there is little difference in cost. However, this remains an interesting optimization for plants where recycling costs are low and/or polymer costs are high.

7.1.2.4 Use of biogas or natural gas for thermal hydrolysis energy requirements

The following looks at the decision to use natural gas or biogas for thermal hydrolysis.

Question. *There is a possibility that biogas could be used directly to provide the energy required for thermal hydrolysis, rather than be used for production of renewable energy. Calculate the impact of doing this on operating costs?*

Answer.
From earlier, the energy required for thermal hydrolysis is 75,700 kWhr/d

This requires 75,700 kWhr/d/10.7 kWhr/m^3 natural gas/85% boiler efficiency

= 8,323 m^3/d natural gas

This costs £2,663/d.

The revenue from renewable energy per day = 3,510 kW × 24 hr/d × £0.14/kWhr = £11,793.60/d.

Therefore, net spent is £2,663 – £11,793.60 = – £9,130.60

Therefore, the calorific value of biogas entering the co-generation plant

= 267,375 kWhr – 75,700 kWhr = 191,675 kWhr

Based on the assumptions surrounding co-generation, this will result in a renewable energy generation of 2,515 kW. This is a loss of approximately 1 MW.

The value of this energy = 2,515 kW × 24 hr/d × £0.14/kWhr = £8,450.40/d.

In this instance, diverting biogas to the boiler has eliminated the daily cost of auxiliary fuel (£2,663) but it has resulted in the loss of (£9,130.60 – £8,450.40=) £680.20 of revenue from biogas/d. Here, it makes financial sense to use biogas for steam production, however, it is dependent on the value assumed for renewable energy. In many instances, it is financially damaging to divert biogas to a boiler, however, it remains the less environmentally viable option.

7.1.2.5 Influence of treating only biological sludge

There are several benefits of only treating biological sludge with thermal hydrolysis as previously described in Chapter 2. However, these benefits come at the expense of providing enhanced treated biosolids (Chapter 5). From a financial perspective, plants treating only biological sludge are smaller, consume less energy and provide similar benefits to those processing both primary and biological sludge.

The following charts show the financial impact of treating only biological sludge based on the following assumptions:

- *Biological sludge is 40% of the feed material*
- *Loading rate for thermal hydrolysis is 5.0 kg VS/m^3*
- *Digestion performance with thermal hydrolysis = 55%*
- *Dewatering performance drops to 30%*
- *Cake recycling costs for both digestion and thermal hydrolysis are £40/t wet*
- *All other assumptions as for previous example*

The changes are highlighted in Table 7.2.

Figures in bold in Table 7.2 are different to data presented in Table 7.1 earlier. The breakdown of operating costs while treating only biological sludge are shown in Figure 7.5. As expected, the main differences relate to the size of the thermal hydrolysis facility and the energy requirements for steam generation. Typically,

Table 7.2 Comparing standard digestion to thermal hydrolysis of biological sludge fraction.

Item	Unit	Digestion Option	Thermal Hydrolysis Option
Electricity for thickening	kWhr/d	5,500	5,500
Polymer for thickening	kg/d	300	400
Anaerobic digestion requirement	m^3	30,000	**15,000**
Electricity for digester mixing	kWhr/d	4,320	1,843
Initial energy required for TH* steam	kWhr/d		**30,282**
Electricity required for TH ancillaries	kWhr/d		**600**
Biogas production	m^3/d	30,375	**37,125**
Biogas yield	m^3/t DS_{fed}	304	**371**
Energy in biogas prior to co-generation	kWhr/d	194,063	**237,188**
Energy required to clean biogas	kWHr/d	7897.5	**9,653**
Renewable energy production	MW_e	2.55	**3.11**
Outstanding energy required for TH steam after recovery of high-grade heat	kWhr/d		**0**
Natural gas required for TH after high-grade heat recovered	m^3/d		**0**
Electricity required to run co-generation plant	kWhr/d	3,056	**3,736**
Electricity for dewatering	kWhr/d	1,800	900
Polymer for dewatering	kg/d	596.25	**587.5**
Nitrogen released requiring treatment	kg/d	1,890	**2,310**
Biosolids cake produced	wet t/d	286	**186**

with systems processing only biological sludge, there is more high-grade heat available than required, therefore, auxiliary gas requirements are eliminated.

Table 7.3 shows the operating cost differences between treating all, or only biological sludge. The main difference in this analysis is that the costs for cake recycling are higher when only biological sludge is processed as the material quality is degraded on mixing with unprocessed primary sludge in the digester, and therefore assumes a higher unit processing cost.

7.1.2.6 Impact on running costs of a dryer

The following example, an extension of the original question shows the importance of thermal hydrolysis in the costs of running a drying plant.

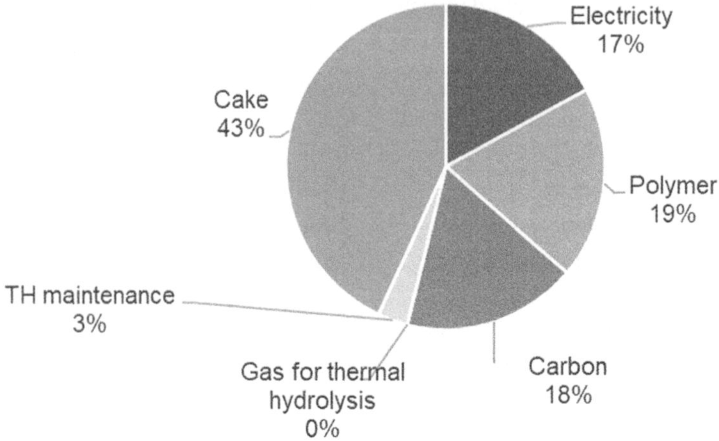

Figure 7.5 Breakdown of operating costs for thermal hydrolysis plant treating only biological sludge.

Table 7.3 Difference in annual running costs between the two previous options and this one processing only biological sludge.

	Anaerobic Digestion	**Thermal Hydrolysis All Sludge**	**Thermal Hydrolysis Only Biological Sludge**
Electricity	£775,400	£880,200	£813,300
Polymer	£817,900	£853,200	£901,100
Carbon	£683,000	£941,000	£834,800
Gas for thermal hydrolysis	£-	£416,000	£-
TH maintenance	£-	£224,700	£143,300
Cake	£4,176,800	£1,739,200	£2,716,300
Costs	**£6,453,000**	**£5,054,200**	**£5,408,600**
Minus			
Biogas benefit	£3,123,800	£4,303,900	£3,817,900
=			
Total Operating Cost	**£3,329,200**	**£750,300**	**£1,590,700**

Question. *For the previous example, the owner wishes to install a drying plant downstream of thermal hydrolysis to dry dewatered biosolids cake to 90% dry solids. Determine the impact of doing this on both capital and operating costs of the system? Assume performance of digestion, thermal hydrolysis and* *dewatering are as for the previous example, in addition to the following assumptions:*

- Capital costs of drying plant are assumed to be £1,500,000 /t water evaporation
- Cake is dried to 90% dry solids
- Thermal energy requirements of dryer are 0.95 MWhr/t water evaporation
- Electricity requirements of running the dryer are 80 kWhr/t water evaporation
- Energy for drying is provided by natural gas

Answer.
Size of drying plant

For anaerobic digestion option:

From before, 286 wet t/cake are produced daily containing 62.94* dry t.
*(=66.25 dry t from digester × 95% capture rate) dry mass.

This contains 286 wet t − 62.94 dry t = 223.06 t water.
If material is dried to 90%, then there are 62.94/90% = 69.93 t pellets.
These pellets contain 6.99 t water.
The difference in water content between the cake and the pellets is therefore

= 223.06 t water − 6.99 t water
= 216.07 t water/d or 9.01 t water/hr.

Therefore, the cost of the dryer in this instance is

= £1,500,000/t water evaporation/hr × 9 t water/hr
= £13,500,000

Repeating the calculations for the thermal hydrolysis:

The cake contains 159 wet t − 50.82 dry t = 108.18 t water.
If the material is dried to 90%, then there are 50.82/90% = 56.5 t pellets.
These pellets contain 5.68 t water.
The difference in water content between the cake and the pellets is therefore

= 108.18 t water − 5.68 t water
= 102.5 t water/d or 4.27 t water/hr.

Therefore, the cost of the dryer in this instance is

= £1,500,000/t water evaporation/hr × 4.27 t water/hr
= £6,405,000

Here it is clear to see the impact thermal hydrolysis has on drying. In this example, the water evaporation, and cost of dryer, is reduced by half. This is typical of full-scale operation. Refer also to Chapter 5, Table 5.4. As with anaerobic digestion, the capacity of existing drying plants can be significantly increased, or size of the new drying facilities reduced.

Operating costs of the drying plant:

These are determined from the quantity of water evaporation required.

For anaerobic digestion option, based on the assumptions:

Thermal energy required/d

$= 216.07$ t water/d $\times$ 0.95 MWhr/t water evaporation

$= 205.26$ MWhr/d $= 205{,}266$ kWhr/d

If biogas is not used, then

$= 205{,}266$ kWhr/d/10.7 kWhr/m^3 natural gas

$= 19{,}184$ m^3 natural gas/d are required before losses

Boiler is 85% efficient, so actual gas requirement

$= 19{,}184$ m^3 natural gas/d/85% efficiency

$= 22{,}569$ m^3 natural gas/d

= £7,222/d

Electricity requirements to run dryer daily

$= 216.07$ t water/d $\times$ 80 kWhr/t water evaporation

$= 17{,}285$ kWhr/d

= £1,383/d

For thermal hydrolysis

Thermal energy required/d

$= 102.5$ t water/d $\times$ 0.95 MWhr/t water evaporation

$= 93.375$ MWhr/d $= 97{,}375$ kWhr/d

If biogas is not used, then

$= 97{,}375$ kWhr/d/10.7 kWhr/m^3 natural gas

$= 9{,}100$ m^3natural gas/d are required before losses

Boiler is 85% efficient, so actual gas requirement

$= 9{,}100\ m^3$ natural gas/d/85% efficiency

$= 10{,}706\ m^3$ natural gas/d

= £3,426/d

Electricity requirements to run dryer daily

= 102.5 t water/d × 80 kWhr/t water evaporation

= 8,200 kWhr/d

= £656/d

The annual operating costs are shown in Table 7.4 for the two options and one where only biological sludge is processed.

A breakdown for the three options is shown in Figure 7.6.

The figure shows the influence of gas costs for drying. If liquor treatment is based on nitrification, then the gas requirements account for half of total operating costs when thermal hydrolysis is absent. This is typical for biosolids drying plants (Sioen & Vriens, 2009), and highlights their dependence on fossil fuels.

Now consider an example where the owner chooses to divert biogas into the drying facility in preference to a co-generation plant.

Table 7.4 Difference in annual running costs between the two previous options and this one processing only biological sludge.

	Anaerobic Digestion	Thermal Hydrolysis All Sludge	Thermal Hydrolysis only Biological Sludge
Electricity	£1,280,400	£1,119,300	£1,103,000
Polymer	£817,900	£853,200	£901,100
Carbon	£683,000	£941,000	£834,800
Gas for thermal hydrolysis	£-	£416,000	£-
TH maintenance	£-	£224,700	£143,300
Pellets	£1,021,000	£824,500	£905,500
Gas for drying	£2,637,100	£1,248,800	£1,513,200
Costs	**£6,439,200**	**£5,627,400**	**£5,400,700**
Minus			
Biogas benefit	£3,123,800	£4,303,900	£3,817,900
=			
Total Operating Cost	**£3,315,400**	**£1,323,500**	**£1,582,800**

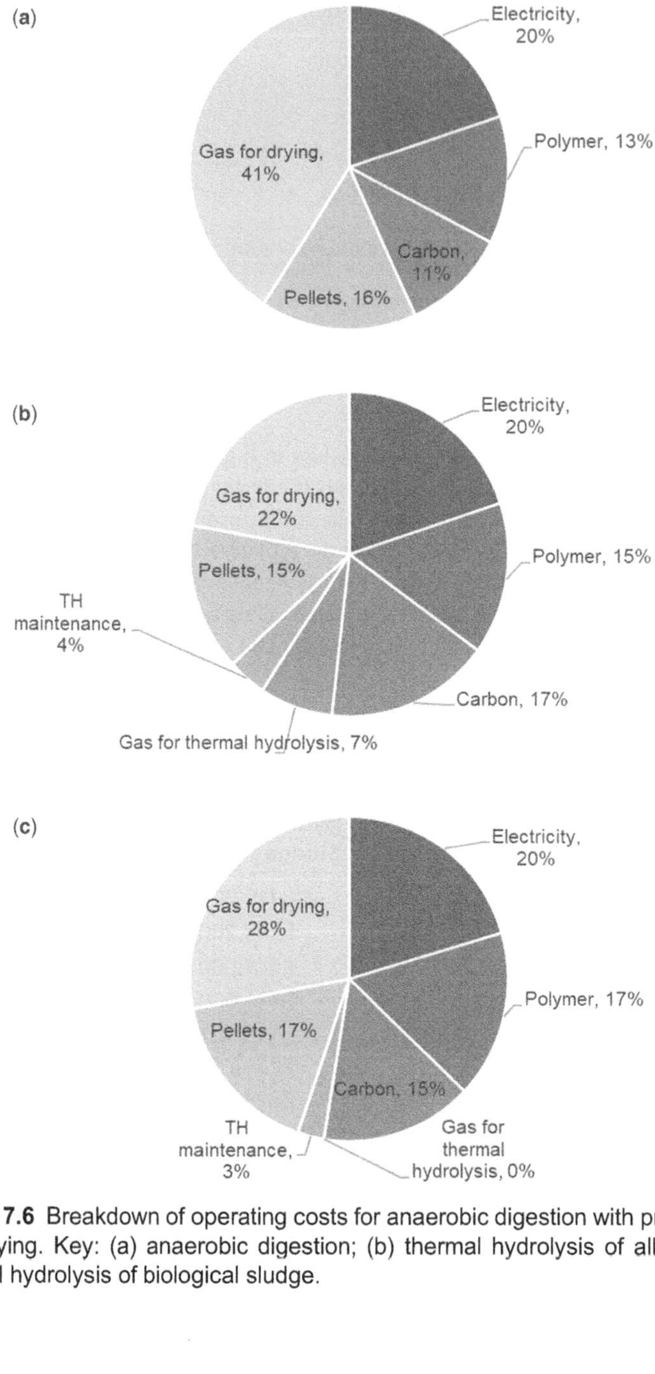

Figure 7.6 Breakdown of operating costs for anaerobic digestion with pre-treatment and drying. Key: (a) anaerobic digestion; (b) thermal hydrolysis of all sludge; (c) thermal hydrolysis of biological sludge.

Question. *How much renewable energy can be made from surplus gas after the requirements of the dryer are met? How does this influence operating costs?*

Answer.
Anaerobic digestion with no pre-treatment

From before, the dryer requires 205,266 kWhr/d energy, and biogas calorific value is 194,063 kWhr/d.

If biogas is used for drying, then the dryer now requires:

$= 205{,}266\ \text{kWhr/d} - 194{,}063\ \text{kWhr/d} = 11{,}203\ \text{kWhr/d}$

$= 11{,}203\ \text{kWhr/d}/10.7\ \text{kWhr/m}^3\text{natural gas} = 1{,}047\ \text{m}^3\text{natural gas}$

$= 1{,}047\ \text{m}^3\ \text{natural gas}/85\%\ \text{boiler efficiency}$

$= 1{,}231\ \text{m}^3\ \text{natural gas which costs} = £394/\text{d}$

This has saved £7,222/d – £394/d = £6,828/d in gas costs for the dryer.
However, this has come at the expense of 2.55 MW of renewable energy generation.
In one day, the revenue from energy generation is

$= 2.55\ \text{MW} \times 1{,}000\ \text{kW} \times 24\ \text{hr/d} \times £0.14/\text{kWhr}$

$= £8{,}568/\text{d}$

Therefore, by using biogas in the dryer, operating costs have increased by £1,740/d

Now with thermal hydrolysis

In this instance, the dryer requires 97,375 kWhr/d energy, and biogas calorific value is 267,375 kWhr/d.
The biogas produced has more energy than required by the dryer, therefore, the dryer energy demands are met outright. Therefore, drying costs of £3,426/d have been avoided.

Now the remaining energy available for co-generation is

$= 267{,}375\ \text{kWhr/d} - 97{,}375\ \text{kWhr/d} = 170{,}000\ \text{kWhr/d}$

Based on the previous assumptions this produces 2.23 MW renewable energy

$(= 170{,}000 \times 35\%\ \text{electrical efficiency of engine}$

$\times 90\%\ \text{availability}/1{,}000\ \text{kW}/24\ \text{hr/d})$

Prior to diverting biogas to the dryer, the plant generated 3.51 MW electricity, therefore, the difference, 1.28 MW has been lost to the dryer.

The value of this biogas is

$$= 1.28\ \text{MW} \times 1{,}000\ \text{kW} \times 24\ \text{hr/d} \times £0.14/\text{kWhr}$$
$$= £4{,}301/\text{d}$$

Subsequently, £3,426/d of drying savings have been offset by loss of £4,301/d revenue from renewable energy, and operating costs have therefore increased by £875/d.

This increase is significantly lower than that incurred by the option where thermal hydrolysis was absent. More importantly with respect to long-term costs, with thermal hydrolysis, the dryer has become independent of natural gas, and therefore is free from long-term fluctuations in the cost of natural gas.

7.1.3 Overall comments on costs of thermal hydrolysis

Thermal hydrolysis may or may not have an attractive payback time. This is critically dependent on the amount of additional infrastructure required, the cost of building new anaerobic digestion capacity, the cost of dealing with the biosolids produced and incentives surrounding the production of renewable energy. However, once installed, thermal hydrolysis can significantly reduce overall operating costs, primarily due to a decrease in biosolids produced—a combination of improved digestion performance with improved dewatering—and increased revenue from renewable energy from biogas. Although more complex than anaerobic digestion without pretreatment, and costs made up of more variables, the benefits from reduced cake and increased energy production far outweigh the increased costs of items like energy for steam, maintenance, and additional liquor treatment.

The analysis presented in this chapter is only to demonstrate the impacts of thermal hydrolysis on costs and is highly dependent on both technical and financial assumptions. Despite this, it shows key findings as follows:

- Thermal hydrolysis reduces the operating costs of anaerobic digestion primarily due to the production of less biosolids, but also the generation of more renewable energy from biogas. These benefits outweigh additional operating costs such as running and maintaining the thermal hydrolysis unit and liquor treatment. The difference in operating costs between digestion with and without thermal hydrolysis widens further when thermal systems are included downstream.
- The cost of liquor treatment can be significant and is important to understand when determining the whole life costs of running a thermal hydrolysis plant. If total nitrogen removal is required, from a strictly cost perspective, deammonification-based systems appear attractive. In this chapter, there has been no mention of refractory nitrogen and COD entering the effluent (Chapter 3), if there are concerns with those, additional infrastructure and associated costs would have to be accounted for.

- There appears little financial benefit to increasing dry solids in the thermal hydrolysis system. The costs saved from reduced energy requirements are offset by increased costs for polymer to reach the higher dry solids required. This is due to the relatively cheap cost of natural gas vis-à-vis polymer. After cake recycling costs and electricity, the polymer is the most significant operating expense of running a thermal hydrolysis plant.
- From an economic perspective, especially when value is given to renewable energy, it is more cost effective to use natural gas, rather than biogas, to meet the energy requirements of thermal hydrolysis. However, this may not be the case from an environmental perspective.
- Generally, processing only biological sludge will increase operating costs but will reduce capital costs. Although the thermal hydrolysis unit is smaller and no steam energy is required, the lack of advanced treatment increases biosolids recycling costs.
- In instances with lower biosolids recycling costs, there may be a cost benefit in intentionally detuning the performance of dewatering equipment to save on polymer costs.
- Thermal hydrolysis significantly reduces the costs of drying (or other thermal processes). This is due to a combination of increased biogas production and less digested biosolids with less water. Drying plants are typically half the size or smaller with thermal hydrolysis than without. It is possible to use thermal hydrolysis with drying to make a drying plant that is independent of fossil fuels and their volatility.

REFERENCES

Abu-Orf M. and Goss T. (2012). Comparing thermal hydrolysis processes (CAMBI™ and EXELYS™) for solids pretreatmet prior to anaerobic digestion. *Proceedings of the Water Environment Federation*, **2012**(2), 1024–1036.

Batstone D. J., Keller J. and Darvodelsky P. S. (2008). Trends in Biosolids Handling Technologies: Economics and Environmental Factors. AWA Biosolids Conference, 7–10 June, Adelaide.

Childress B., Sullivan K., Hanna M., Greenyer J., Cummings P., Lyon-Galvin A. S., Crayden T. and Taylor D. (2019). Biosolids system improvements pay for themselves. *WE&T Magazine*, **31**, 27–31.

Fountain P. (2019). Anaerobic Digestion and Advanced Anaerobic Digestion, Why? A UK Wastewater Perspective, Presentation given at Global Water Summit.

Goss, (2015). Lessons Learned from Recent Thermal Hydrolysis Plants Start up and Commissioning. Presented at WEAT-NTS February Seminar.

Goss T., Stitt B., Moncholi M. and Diaz I. (2017). Centrifuge Piloting at Miami's Central District WWTP–Overcoming obstacles with a difficult to handle sludge. *Proceedings of the Water Environment Federation*, **2017**(6), 4491–4508.

Jolly M. and Gillard J. (2009). The Economics of Advanced Digestion. Proceedings of the 14th European Biosolids and Organic Resources Conference and Exhibition, UK.

Koodie A., Walker C. and Strange G. (2017). PR19 Investment Decisions: Building up a cost for a Thermal Hydrolysis Facility. Proceedings of European Biosolids and Organic Resources Conference, Leeds, UK.

Loomis P., Jacobs T., Mathur S., Guven E., Hurtado D. and Christy P. (2016). Thermal Hydrolysis – The Next Generation. *Proceedings of the Water Environment Federation*, **2016**(3), 903–914.

Moncholi M., Goss T., Stitt B. and Diaz I. (2018). Miami-Dade water and sewer sludge thickening and dewatering centrifuge pilot studies at two of the Southeast's largest wastewater treatment plants, *Florida Water Resources Journal*, **69**, 30–53.

Sioen H. and Vriens L. (2009). Improving the Energy Efficiency of Indirect Sludge Drying. Proceedings of the 14th European Biosolids and Organic Resources Conference and Exhibition, Leeds, UK.

Smith R. W. and Rus E. (2019). Techno-Economic Analysis of Anaerobic Digestion Pre-Treatment Technologies for a UK Site. Proceedings of WEF/IWA Residuals and Biosolids Conference, Ft. Lauderdale, Florida, pp. 1197–1202.

Van Horn. (2015). Thermal Hydrolysis: The Ins and Outs, Ups and Down. Presentation at Ohio WEA Biosolids Conference.

Urban Wastewater Treatment Directive – Council Directive 91/271/EEC of 21 May 1991.

Williams T. (2014). Why Thermal Hydrolysis with Anaerobic Digestion is Rising to the Top in North America. Presented at the Proceedings of 19th European Biosolids and Organic Resources Conference and Exhibition, Manchester, UK.

Willis J., DeClippeleir H., Graf W., Kumar A., Lucas B., Murthy S. and Sharma K. (2015). DC Water's sewer-methane carbon footprint. *Proceedings of the Water Environment Federation*, **2015**(6), 6228–6239.

Chapter 8

Future developments

Those of us who have been interested in sewage purification for many years constantly, as Kipling remarked in regard to Homer's listeners, 'hear old songs come round again' simply dressed up in new words; that is to say, much of the research of past years is rediscovered by new workers and given to us again

H. W. Clark – In *Sewage Works Journal*, Vol. 2, No. 4, 1930

8.1 INTRODUCTION

Although this chapter refers to potential new uses and directions for thermal hydrolysis, as the quote above states, "new" ideas and concepts typically find their routes in the past. For example, the use of thermal hydrolysis for hydrothermal carbonization, proposed later in this chapter, is based on technology from the 1930s, whereas use of advanced anaerobic digestion systems combined with thermal hydrolysis is based on digester designs from the late 1960s and 1970s. The present approaches to wastewater treatment are founded on technology which is over 100 years old, in response to drivers of that time. Subsequently, existing processes and technology will struggle to be sustainable in future years, and technologies which optimize these, such as thermal hydrolysis, have a place in the near to medium future to assist with meeting current drivers. However, moving longer term, fundamental changes are required in the way wastewater and sludge are treated. There is currently a great deal of research looking forward, and the vast majority of this is originating from China. Research from other countries, where perhaps thermal hydrolysis has been established

Author: William Barber
doi: 10.2166/9781789060287_0215

longer, typically involves looking at specific operational concerns and optimizations, such as investigating nutrient balances and addressing impacts on liquor treatment. In this chapter, potential future avenues for the process are divided into three areas: (1) further optimizing of anaerobic digestion; (2) generation of different products and (3) sterilization.

8.2 FURTHER OPTIMIZING THE DIGESTION OF SEWAGE SLUDGE

There are many potential avenues for the use of thermal hydrolysis moving forward with respect to improving anaerobic digestion. However, these designs would need to be accompanied by different digestion configurations; otherwise the digestion process itself will remain limited in performance. Designs to date remain conservative, and there is data to suggest that lower retention times are feasible – in fact, thermal hydrolysis is better suited to lower retention times – and it is possible to digest at higher dry solids (DSs), especially when the sludge mix is laden with primary sludge, due to a combination of lower viscosity and lower nutrient content.

8.2.1 Plug-flow digestion

With upstream configurations, thermal hydrolysis can improve volatile solids destruction up to approximately 65%. However, this still leaves a third of the volatile material and energy within the treated biosolids. With downstream configurations, volatile solids destruction can increase further to 70–75%, but this comes at the expense of losing enhanced treated biosolids status. On plants with multiple digesters, an easy way to do this is by installing thermal hydrolysis prior to anaerobic digestion which is run in series, not parallel. Doing this causes a shift in bacterial populations allowing increased protection against toxic materials and higher resistance to changes in environmental parameters such as pH and temperature (Barber & Stuckey, 1999). Probably greatest advantage of running digesters in this configuration is its ability to separate acidogenesis and methanogenesis across the reactor system to make it behave like a two-stage system (Weiland and Rozzi, 1991). Two-phase operation can increase acidogenic and methanogenic activity by a factor of up to 4 as acidogenic bacteria accumulate within the first stage and different bacterial groups can develop under more favorable conditions in latter reactors (Cohen *et al.*, 1980, 1982). By increasing reaction kinetics, thermal hydrolysis would be complementary to series operation of digesters. Furthermore, as more acids will be produced in the first digester, this will lower system pH which will provide stability by reducing free ammonia concentrations when combined with thermal hydrolysis. In a more recent work, Chapman and Muller (2010) showed that a medium performing plant at Port Angeles (Washington, USA) increased in volatile solids destruction from 38% to 62% when switched over from parallel to series operation. Even on other facilities which performed well with volatile solids destruction in the range of 60 to 65% still improved by 5 percentage points to 70% destruction. Series operation may explain the superior performance of intermittent thermal hydrolysis over that installed upstream. Attempts at modelling series digestion often underpredict performance improvements due to a lack of kinetic data for true plug-flow systems and layers of conservative assumptions. Subsequently, models may predict modest improvements (between 5% and 10% biogas increase); however, full-scale plants which have been converted from parallel to serial-fed ones have seen biogas increases between 20% and 40% (Chapman & Muller, 2010).

Long Reach facility owned by Thames Water has series digestion following thermal hydrolysis. The biogas yield from the plant is in the region of 400–440 $m^3/t\ DS_{fed}$, which is similar to other facilities owned by Thames water in spite of only the waste activated sludge being processed (Rus *et al.*, 2017).

8.2.2 Advanced digestion designs

Qiao and colleagues (2013) digested thermally hydrolyzed sludge liquors using a mesophilic granular sludge bed bioreactor at loading rates of 11 kg COD/m^3 d. This loading rate is approximately 5.8 kg VS/m^3 d, which does not appear overly stressful for this type of anaerobic digester. However, the reactor managed to achieve a VS destruction rate of between 60% and 65% while operating at a retention time of 2.5 d. Energy produced from biogas could meet significant requirement of the required steam energy. The intent of the work was to combine maximized biogas production in short retention time, and produce biosolids which could be co-incinerated with municipal solids waste. Total energy production from combined biogas and incineration systems was approximately 13,500 kJ/dry t processed while the energy demand for thermal hydrolysis was in the region of 1,900 kJ.

In an earlier work, the same research group investigated the biogas potential from hydrolyzate in an Upflow Anaerobic Sludge Blanket (UASB) reactor (Qiao *et al.*, 2011). In this instance, thermal hydrolysis operating conditions were 170°C for 60 min. The authors took sludge cake from a full-scale plant and diluted it to 10% (to assist with pumping under laboratory conditions) prior to thermal hydrolysis. Afterwards, the hydrolyzed slurry was dewatered again and the supernatant fed to a UASB reactor. The supernatant contained 25,000, 2,700 and 710 mg/l COD, total nitrogen and total phosphorous, respectively. Loading rates up to 24 kg $COD/m^3/d$ (equivalent to approximately 16 kg $VS/m^3/d$) were successful with removing most of the COD, with upflow velocities up to 1 m/hr. Granulation was observed after 150 d of operation and had a smooth shape with diameter between 1 and 3 mm. Granulation significantly increased the specific methanogenic activity of the sludge, as shown in Table 8.1 taken from Qiao *et al.*, 2011).

Table 8.1 Performance of UASB treating supernatant from thermally hydrolyzed municipal sludge (from Qiao *et al.*, 2011).

OLR (kg COD/m^3/d)	Specific Methanogenic Activity (g COD CH_4/g VSS/d)	COD Removal Rate (%)
24.2	6.46	64
16.1	5.17	81
17.3	14.45	72

> *Loading rates of up to 24 kg COD/m^3 digester volume/d are possible with thermal hydrolysis and high rate digestion systems*

Previously, an anaerobic sequencing batch reactor (ASBR) had been used successfully with thermally hydrolyzed sewage sludge (Wang *et al.*, 2009). The ASBR operates in a similar way to a traditional aerobic sequencing batch reactor in a cyclic batch mode with feeding, reaction, thickening and decanting cycles. The authors noted that this type of anaerobic digester is well complemented by thermal hydrolysis, due to the latter technology making sludge far more settleable thereby enhancing the thickening and decanting cycles. Furthermore, the release of COD improves the reaction cycle. Wang's co-workers compared the performance of an ASBR against a standard anaerobic digester processing thermally hydrolyzed sludge at 10 and 20 d HRT with corresponding loading rates of 5.4 and 2.7 kg COD/m^3/d. At 20 d, biogas production was 15% higher in the ASBR, but due to faster kinetic rates, gas production was over 30% higher at the lower retention time. These data were derived from volatile solids reductions of 54 versus 64%, and 45 versus 56% for control and ASBR at 20 and 10 d, respectively. Interestingly, concentrations of soluble COD, alkalinity and acetic acid were similar for both reactor designs. Solids concentration increased in the ASBR overtime (between 65 and 80 g/l) resulting in a solids retention time of close to 40 d while maintaining a hydraulic retention time of 10 d. However, solids concentration accumulated to a point where the ASBR became unstable as the solids had consumed reactor space and further lowered the hydraulic retention time, highlighting the importance of occasional solids wastage with this reactor design.

8.2.3 Recuperative thickening with thermal hydrolysis

In a recent work, Li's group has investigated the use of recuperative thickening to separate solids from hydraulic retention times (Li *et al.*, 2019) treating hydrolyzed sludge. Besides keeping solids within the reactor, it was thought that recuperative thickening would provide sufficient time to break down poorly degradable substrates. The study showed that increasing retention time from 15 to 30 d had no impact on digestion, either positive or negative. This implied that the material which hadn't degraded was refractory and that, in theory, it is possible to use recuperative thickening to further increase the loading rate of digesters feeding on hydrolyzed sludge. It was possible to maintain an HRT of 6–7 d which corresponded to a solids retention time of 15 d with no detrimental impact. The results highlighted that anaerobic digestion can be further exploited with thermal hydrolysis than current full-scale designs would suggest. However, work on microbiological analysis has shown that many of the trophic groups which specialize in aiding hydrolysis reactions are in negligible numbers in digesters fed thermally hydrolyzed material, and this needs to be accounted for when trying to improve hydrolysis rates during digestion.

8.2.4 Higher dry solids

Presently, digesters fed hydrolyzed sludge are limited by the concentration of free ammonia in the digester more than they are the viscosity of the incoming sludge. Subsequently, this sets loading rates (Table 5.1, Chapter 5) and for mixed sludge, these are equivalent to feeding anaerobic digesters at approximately 10% DSs. This is achieved by addition of dilution water as described in Chapter 2. However, the nitrogen content as well as viscosity is fundamentally different for primary and biological sludge. Based on digestion of biological sludge exclusively, to achieve similar nitrogen concentrations in the digester (based on the approach used in Chapter 4), DSs fed would have to reduce to approximately 7.5% DSs. Conversely, if the sludge was exclusively primary, DSs could be increased to circa 16% DSs which is potentially where dilution is not necessarily making viscosity the limiting factor. Furthermore, data have shown that digesters can become acclimated to higher nitrogen content (as described in Chapter 3). In Brussels, a commercial-scale thermal hydrolysis plant is followed by both anaerobic digestion and sub-critical wet air oxidation, where the sludge is routinely fed at 13–14% DSs with no detrimental impact on the performance of downstream operations. An alternative to acclimation to higher ammonia concentrations is the use of ammonia stripping.

8.2.5 Ammonia stripping

Stripping of ammonia to reduce toxicity and therefore improve digestion performance has been used successfully on a variety of digestion systems. Typically, these are based on extracting a sidestream, manipulating temperature and pH to encourage a shift in equilibrium toward free ammonia, and then using a fluid to strip the ammonia from the aqueous phase. Bank's team in Southampton (Banks *et al.*, 2016) looked at stripping ammonia to improve the digestion of a high protein foodwaste under thermophilic conditions at pilot-scale. Biogas was used to strip ammonia under different operating conditions (55–85°C, pH 10 and unadjusted, and stripping times of 2–5 d). By using a temperature of

>70°C at pH 10, almost 50% of the total ammonia nitrogen was stripped out which provided digester stability. This was confirmed when the stripping was stopped on one of the test digesters which resulted in an increase in ammonia and ultimate failure.

More recently, ammonia stripping has been applied to thermal hydrolysis of digested sludge (Yang *et al.*, 2019). That study directly compared a control – based on standard mesophilic digestion – with thermal hydrolysis (160°C for 30 min) in both anterior and posterior positions with respect to mesophilic digestion. With the positioning downstream of digestion, centrate, which would normally be sent back for digestion (Chapter 2) was stripped of ammonia prior to recycle. As with the work of Bank's group, Yang's team adjusted the centrate pH to 10 and the temperature to 70°C. However, in this instance, air, rather than biogas was used for stripping. Total ammonia nitrogen was kept below 500 mg/l in the stripped centrate. Tests showed that pre- and post-addition of thermal hydrolysis increased biogas yield compared to a control by 30% and over 40%, respectively. Interestingly, the authors noticed that the biogas yield of raw sludge plus the treated returned hydrolyzed centrate was higher than that of the raw sludge plus a measurement of the biogas potential from the treated centrate when digested separately. This suggests a level of synergism when operating in a post-digestion configuration. The authors noted that the digestion system with ammonia stripped centrate return showed high levels of stability.

8.2.6 Lower hydraulic retention time

As thermal hydrolysis acts to enhance hydrolysis, it is possible to accelerate the rate of gas production, such that the same amount of biogas can be produced in a shorter time. The work of Xue's group (2015) shows this influence very clearly in a graphical format. In their thorough study which looked at reaction temperatures between 60°C and 180°C and times between 15 and 180 min, the value of accelerated biogas production is optimal at approximately 10 d. Observing data taken at a reaction temperature of 140°C, biogas production was approximately 70% higher with thermal hydrolysis than without at a reduced retention time of 10 d. When retention time was increased to 20 d, the difference dropped to <25% additional biogas for the hydrolyzed system, as the biogas production from the control sluggishly narrowed the gap. Interestingly, in the control, the biogas yield at 10 d was approximately two-thirds of that at 20 d, while the reactor with pre-treatment was producing almost 95% of the 20 d yield within 10 d. This finding is mirrored by Ngwenya's team (2015) who showed statistically insignificant difference in biogas production with thermal hydrolysis between 10 and 18 d at equivalent loading rates at the laboratory scale. Their work, along with (Bougrier *et al.*, 2008) implies that carbohydrates are degraded prior to proteins during digestion (regardless of thermal hydrolysis), such that ammonia and alkalinity and consequentially pH increase with retention time, increasing the potential for free ammonia inhibition. As the production of extracellular microbial byproducts also increases with retention time (Barker & Stuckey, 1999) with a subsequent increase in viscosity influencing dewaterability (Neyens & Baeyens, 2003) it may be hypothesized that running digestion plants at approximately 10 rather than 20 d retention time may be preferable when coupled with thermal hydrolysis. Li and Noike (1992) concluded optimum digestion retention times between 5 and 10 d based on various tests and observations of changes in methanogenic populations. Chertsey, in the UK, has been running at hydraulic retention time of approximately 12 d at full-scale with good performance (Pook *et al.*, 2013; Walley, 2007) while Wilson's team (2008) concluded that digestion performance at 15 d retention time with thermal hydrolysis was equivalent to that without thermal hydrolysis at 20 d.

8.2.7 Digestion of algae

There has been growing interest in recent years in the exploitation of algae for wastewater treatment as a means to remove nutrients (Mallick, 2002); however, intentional growth of algae to generate renewable energy through anaerobic digestion has been studied since the mid-1970s in response to the oil crisis (Chynoweth *et al.*, 1980). Recent studies have shown that autoclaving *Sargassum* sp. at 121°C and 1 bar for 15 min increased the soluble VS content by a factor of 10 resulting in a 60% increase in biogas yield compared to a control where the algae was unprocessed (Costa *et al.*, 2015). Hydrothermal pre-treatment at 130°C and 160°C with steam explosion has been very successful for improving the degradability of *Saccharina latissima* with a reaction temperature of 10 min (Thompson *et al.*, 2019). Biogas production increased by up to 20% compared to a control without thermal hydrolysis.

8.2.8 Chemically enhanced thermal hydrolysis

Rather than being an alternative process, thermal hydrolysis complements a variety of other technologies. As explained earlier, increasing reaction temperature improves biogas production up to a certain point (dependent on sludge type) above which the production of recalcitrant components reduces digestion performance. In contrast, dewaterability improvements appear to continue to rise with temperature, due to the destruction of extracellular polymer material which makes the sludge more compressible. Potentially, chemicals can be utilized prior to thermal treatment to start the lysis process and thereby reduce baseline viscosity. This chemically lyzed material can then be heat treated to temperatures below which refractory compounds are created in bulk, and therefore provide the benefits of thermal hydrolysis with reduced risk due to chromophore production. This approach is gaining increased traction amongst researchers in China, although it has been looked at for over 15 years (Valo *et al.*, 2004). In Valo's work, thermal hydrolysis temperatures between 130°C and 170°C were tested with and without hydrogen peroxide and Fenton's reagent pre-treatments. The authors found that solubilization of COD increased with pH (up to 12), and this was approximately three times the level observed by chemical addition at ambient temperature. However, COD solubilization did not reveal organic matter solubilization as volatile solids measurements remained uniform. Laboratory-scale digesters revealed that 130°C and pH = 10 was more efficient at enhancing methanogenesis than processing at 170°C with no chemical addition (Valo *et al.*, 2004). Later work aimed at elucidating the benefits of combining hydrogen peroxide addition with thermal hydrolysis specifically on dewaterability and biogas production (Abelleira *et al.*, 2012). The authors found that even modest additions of hydrogen peroxide prior to thermal hydrolysis resulted in far better dewatering (measured as time to filter) than thermal hydrolysis in the absence of the oxidant, even though thermal hydrolysis was conducted at lower reaction temperature. As with previous authors, Abelleira's coterie found far higher levels of solubilization when chemicals were added prior to thermal hydrolysis. While promising, there are various issues to consider:

- The degradation of chemicals at higher temperature rendering them largely ineffective
- Boiling point of hydrogen peroxide is <150°C
- Risk associated with the use of oxidants at elevated temperatures and pressures
- Health and safety of operators
- Corrosion of metal vessels and pipework

8.2.9 Intermediate options for digestion improvements

There are a number of less complex ways to adjust existing anaerobic digestion plants in order to obtain greater performance out of them, which may provide improvement in the intermediate future on plants with several digesters. These are shown in Figure 8.1.

Figure 8.1a shows a typical parallel-fed digestion plant. Installation of thermal hydrolysis in front of this facility (Figure 8.1b) will enhance performance further as described in Chapter 5. Based on the work by Chapman and Muller (2010) volatile solids destruction will be further enhanced if the digesters are fed in series. In Figure 8.1d the hydraulic retention times for both primary and waste activated sludge are decoupled while maintaining series operation. By reducing retention time for digestion of primary sludge, the biological sludge gains further retention time to produce more biogas. Finally, it is possible to reroute the digested primary sludge into a downstream thermal hydrolysis plant along prior to further digestion with the hydrolyzed biological sludge (Figure 8.1e).

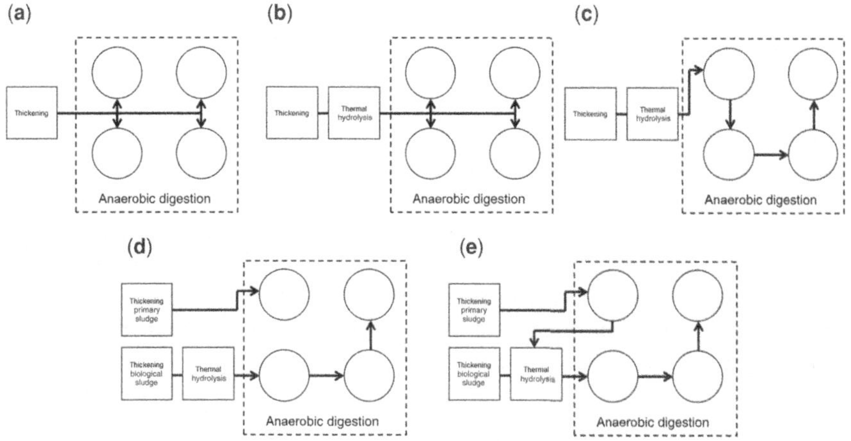

Figure 8.1 Further improving anaerobic digestion with thermal hydrolysis: (a) standard parallel-fed anaerobic digestion; (b) similar to (a) preceded by thermal hydrolysis; (c) similar to (b) with series feeding of digestion; (d) similar to (c) but separate digestion of primary and biological sludge with thermal hydrolysis only of biological sludge and (e) similar to (d) but digested primary sludge fed with biological sludge to thermal hydrolysis prior to further digestion.

8.3 PRODUCT FORMATION

8.3.1 Production of char and chargas from hydrothermal carbonization

Thermal hydrolysis processes can be operated under different conditions to enable hydrothermal carbonization. This is an exothermic process which converts sewage sludge into carbonaceous products principally char, at operating temperatures slightly above those experienced by thermal hydrolysis, typically in the range of 180–250°C. Retention times vary from minutes to hours. The elemental composition of sewage sludge is significantly altered after hydrothermal carbonization, resulting in an increase in carbon with decreases in oxygen, nitrogen and hydrogen (Wang *et al.*, 2019). This increases the calorific value of the char with increasing reaction temperature (Peng *et al.*, 2016). Peng's study measured calorific values of 10.97 and 12.06 MJ/kg dried sludge for sewage sludge and char produced at 260°C for 60 min, respectively. Char yield increases with reaction time but falls with reaction temperature from approximately 65% yield at 180°C to 53% yield at 300°C (Peng *et al.*, 2016). Danso-Boateng and co-workers (2015) reported that the global kinetics of char production from primary sewage sludge at 140–200°C could be predicted through a first-order relationship with an activation energy of 70 kJ/mol.

In addition to the char, the process yields process water which can contain short-chain fatty acids (especially acetic) and other organics including furanic, phenolic, aromatic, alkene and aldehyde compounds (Wang *et al.*, 2019). A carbon dioxide-rich gas is also produced. The concept of hydrothermal carbonization was first described in the 1920s investigating chemical reactions from coal (Bergius, 1932). Similar to thermal hydrolysis, it is well documented that hydrothermal carbonization significantly improves dewatering with improvements greater than those shown for thermal hydrolysis but at higher temperatures. Converting existing thermal hydrolysis equipment and expertise in the field to accommodate the temperatures and pressures required by hydrothermal carbonization and other thermal processes could yield various products in the future. In fact, work is being undertaken in China by Chen *et al.* (2020) which is studying the influence of thermal hydrolysis in combination with pyrolysis.

8.3.2 Generation of proteins and similar products

The materials solubilized by thermal hydrolysis can be extracted and concentrated rather than fed into an anaerobic digestion system. In China, Yuchuan Environment, have been using thermal hydrolysis at temperatures of up to 130°C, combined with alkalis to make a variety of high-value end products from direct dewatering of the hydrolyzate (Wang, 2018). Approximately 1–3 g/l proteins released from hydrolysis are present in the hydrolyzed liquors and these are concentrated by evaporation and used in an assortment of products including: foaming agents; foam fire extinguishers; liquid fertilizers and industrial proteins (Figure 8.2). Alternatively, the liquors can be fed to anaerobic digestion for biogas production. The solid fraction may be dried and used in a range of outlets including soil conditioning, thermal insulation material or organic fertilizers (Wang, 2018). The thermal hydrolysis reaction is between 3 and 4 hr, with steam demand of 0.12–0.2 t/wet t processed (equivalent to 0.6–1 t/dry t). The long reaction time results in less than 100 mg/l suspended solids.

An alternative source of high-value nutrients can be found within the digestate produced. Work in Australia (Ward *et al.*, 2018) is looking at commercializing the potential for recovery of melanoidin products from digestate. At the time of writing, the team is showing that the energy recovered per unit of nitrogen is comparable to other, more traditional approaches for nutrient management.

There is a growing trend to combine dedicated nutrient recovery processes, whereby conditions are altered, and chemicals added to precipitate nutrients such as struvite and brushite, with thermal hydrolysis. The two systems appear well-suited due to concentration of sludge by thickening prior to thermal hydrolysis and increased solubilization of nutrients (as described in Chapter 3). Ideally, the nutrient removal process would be positioned after thermal hydrolysis and prior to subsequent digestion.

Figure 8.2 High value protein product extracted from biosolids which have been treated with alkali and thermal hydrolysis (Photo by author).

8.4 STERILIZATION

8.4.1 Treatment of antibiotic-resistant bacteria and genes

The worldwide increase in antibiotic-resistant bacteria is considered as a major challenge by the World Health Organization (WHO) and was discussed at the G7 summit. Large amounts of antibiotics are released into municipal wastewater due to incomplete metabolism in humans or due to disposal of unused antibiotics. Subsequently, both antibiotic-resistant bacteria and antibiotic-resistant genes (ARBs and ARGs) have been detected in wastewater samples. Data show significantly higher proportion of ARB contained in raw and treated wastewater relative to surface water, and various studies conclude that conditions in wastewater treatment plants are favorable for the proliferation of ARBs and ARGs (Bouki *et al.*, 2013). In addition, several studies indicate that the environmental conditions in waste-water treatment plants may enhance the likelihood of gene transfer to non-resistant bacteria (Davies, 2012). Wastewater treatment plants show varied potential for removal of ARBs and ARGs, although there is still a fundamental lack of data on abundance and fate of these materials. In principle, thermal hydrolysis which operates at temperatures in excess of autoclaves used for sterilization can play an important role in the future with controlling quantities of ARBs and ARGs.

Work in this area is ongoing in China. One study considered the potential of mesophilic digestion by itself or combined with either ozone or thermal hydrolysis pre-treatment on the destruction of five tetracycline-resistant genes (*tet*A, *tet*G, *tet*Q, *tet*W and *tet*X) and one mobile-genetic element (MGE, intI1; Pei *et al.*, 2016). MGEs are sequences of genetic material which change positions on a chromosome, or are exchanged between chromosomes, bacteria or even species. The study compared pharmaceutical waste sludge with that produced from municipal treatment. As would be expected, the raw pharmaceutical wastewater had at least an order of magnitude more *tet* genes than municipal sludge (10^9–10^{13} versus 10^9–10^{10} copies/g DS correspondingly). Thermal hydrolysis by itself resulted in a log reduction of between 2 and 3.8 for the genes. When measurements were taken post digestion as well, *tet*A and *tet*G levels remained at the same level, but the other genes increased significantly, although still remained lower than the raw material for both sludge types. However, the same could not be said for ultrasonic treatment, with digestates containing higher quantities of some of the ARGs. Additional work in China has looked at ARGs and MGE during full-scale operation for two plants in Beijing (Tong *et al.*, 2019). Abundance of ARGs was quantified as 115.7 and 113.1 log copies/g raw sludge for the two plants.

After thermal hydrolysis, ARGs were reduced by 25.5 and 20.8 log and MGE by 4.71 and 3.55 log, respectively for the two installations. However, these removal rates were inconsistent with the earlier work of Pei's group (Pei *et al.*, 2016). Thermal hydrolysis was found to be highly effective at removing: bla_{TEM}, $bla_{\mathrm{NDM\text{-}1}}$ (carbapenem-resistant gene), *erm*B, *erm*F, *mef*A/E, *qnr*A, *qnr*S, *tet*M, *sul* I, *sul* II, *intI*1 and Tn916/1545 but had little influence on $bla_{\mathrm{CTX\text{-}M}}$. However, following subsequent anaerobic digestion, with the exception of bla_{TEM} (removal of which remained high), *qnr*A, *tet*A, *tet*X, *sul* II, *intI*1, the levels of the remaining ARGs and MGEs increased to be higher than the raw sludge data. Previous work has found enrichment of ARGs of erythromycin esterase type I, sul1 and tetM during mesophilic digestion although digestion was found to reduce other materials which were tested (Zhang *et al.*, 2015). Zhang's team showed that efflux pumps were the most common mechanism for antibiotic resistance in sludge. In Tong's work (2019) total relative abundance of ARGs and MGEs was higher in the thermally hydrolyzed digested sludge than they were in the raw feed for one of the sites, while on the other, they were lower. These inconsistent results highlight the need for future research in order to find abatement and treatment strategies for antibiotic resistance (Bouki *et al.*, 2013). Thermal hydrolysis may play an important role in antibiotic resistance either with, or most likely without, ensuing digestion.

8.5 RECOMMENDATIONS AND FUTURE WORK

Thermal hydrolysis has significantly evolved since installation of the first facility in Norway in 1995. Subsequently, it has become a well-established commercially viable technology which is considered a mainstream process in several global geographies. Thermal hydrolysis reduces operating costs and carbon impact compared when benchmarked against other approaches of sludge management. The primary benefit of the technology involves changing sludge rheology which allows higher loading rates to digesters, dryers and incinerators, and significantly improves dewatering irrespective of dewatering device. This results in much less biosolids being produced, and along with a higher level of treatment, opens new agricultural outlets. Although the energy benefit is relatively neutral when parasitic load is accounted for, the energy balance of downstream thermal processes is fundamentally improved due to a combination of requiring less energy for thermal processing and production of more energy within biogas. Newer configurations have been developed which eliminate the parasitic energy requirement, further improving the energy balance. This is fundamentally important moving into the future, as it is possible to have facilities which are energy self-sufficient and therefore not exposed to fluctuations in energy price. Currently, anaerobic digesters preceded by thermal hydrolysis are limited by free ammonia toxicity, however there is an abundance of literature which suggests that this limitation can be overcome, which could see higher loading rates than are presently being adopted. Additionally, evidence suggests that digesters fed thermally hydrolyzed sludge are suited to far lower hydraulic retention times than they are currently being designed for, with times of 10–12 d being sufficient. Longer retention times encourage protein degradation which increases ammonia, alkalinity and pH, and do not result in a statistically significant increase in biogas production. Therefore, designing for lower retention times could not only reduce capital expenditure, but also improve performance within the digester. New work is showing that thermal hydrolysis complements higher rate digestion systems capable of further increasing loading rates. Hydrolysis reaction temperature plays a fundamental role in regulating the efficacy of the overall biosolids treatment plant. Higher temperatures improve dewaterability but come at the expense of refractory compound generation which carry various detrimental impacts if not accommodated for. Effluent treatment requirements may influence the necessity for thermal hydrolysis and must not be overlooked. When looking at co-digestion, higher reaction temperatures are not required, and are in fact detrimental as they encourage melanoidins. Therefore, plants with co-digestion can routinely operate at 145°C or lower and still maintain high biogas production levels. It is proposed that rather than a rigid set of operating conditions, these can be tailored to meet specific site requirements. Thermally hydrolyzed sludge has fundamentally different characteristics to unmanipulated sewage and these need to be accounted for in design of subsequent plant, especially the digestion system, where reduced bulk density and rapid rise events may be influential. Despite the maturity of the technology, there are several outstanding areas which require further research to enable its optimization including:

- A better understanding of the production of refractory compounds and their impacts on downstream processing
- More data on the thermodynamics of the system to enable higher throughputs and greater efficiency
- Further knowledge on the interactions with nutrients such as nitrogen and phosphorous, especially regarding their extraction and development of nutrient products
- More research into other applications of thermal hydrolysis, for instance hydrothermal carbonization
- Better understanding of the performance of combination systems, for instance combining thermal hydrolysis with chemicals, ultrasound, other lysing systems etc.
- More research is needed into understanding the influence of thermal hydrolysis on emerging contaminants and antibiotic resistance in sludge
- It is necessary to develop standardized protocol and testing procedures to enable better cross-referencing between research groups

Table 8.2, adapted from Barber (2016), provides some recommendations for the use of thermal hydrolysis depending on drivers and site requirements

Table 8.2 Recommendations for thermal hydrolysis (adapted from Barber, 2016).

Aim/Concern	Recommendations
Minimize construction of anaerobic digestion plant	• Thermally hydrolyze both primary and activated sludge prior to anaerobic digestion. Design digestion plant at 10–12 d retention time to maximize size reduction • Maximize pressure drop to increase biogas production rate to allow shorter digestion retention times • Install modular plant to reduce on-site construction time
Maximize overall energy balance around digestion	• Thermally hydrolyze only activated sludge. Optimal proportion of primary sludge is approximately 30% of incoming feed • Reduce thermal hydrolysis reaction temperature • Thermally hydrolyze digested sludge prior to a second stage of anaerobic digestion

(*Continued*)

Table 8.2 Recommendations for thermal hydrolysis (adapted from Barber, 2016) (*Continued*).

Aim/Concern	Recommendations
	• Run downstream digestion plant in thermophilic region but no higher than 53°C • Install thermal hydrolysis downstream of digestion and recycle solubilized centrate
Design for best dewatering	• If installed prior to digestion, run thermal hydrolysis plant at highest possible reaction temperature • Install thermal hydrolysis downstream of anaerobic digestion plant immediately prior to dewatering which should be conducted at high temperature • Run thermal hydrolysis plant immediately prior to dewatering in absence of anaerobic digestion
Foaming concerns	• Run thermal hydrolysis reactors at highest temperature to minimize viscosity of sludge and solubility of material, both of which minimize foam stabilization
Concerns with color and refractory material	• Run thermal hydrolysis reactors at lower temperature (preferably <150°C) • Production of refractory material increases proportionally with elevating quantities of activated sludge • Avoid addition of sugary material which encourages melanoidin production by combining with protein in activated sludge • Use coagulants during dewatering • If necessary, install purpose build facility, such as ozonation plant to remove remaining refractory material
Co-digestion	• Run thermal hydrolysis operating temperature approximately 20–30°C lower than for sewage sludge • Care needs to be taken with respect to addition point of organics. If added upstream of thickening, a third to a half of the soluble COD may be lost in the thickening stage and end up as an additional operating cost during biological treatment. If pasteurization is not required, it may be better to add material downstream of thermal hydrolysis prior to digestion
Liquor treatment	• A fraction of the returned COD is bioavailable and can be used for denitrification

(*Continued*)

Table 8.2 Recommendations for thermal hydrolysis (adapted from Barber, 2016) (*Continued*).

Aim/Concern	Recommendations
	• If no spare capacity available in the existing biological treatment plant, deammonification provides a lower operating cost • If deammonification used, it is necessary to pre-treat the returns from dewatering to remove solids and COD. Alternatively, COD can be diluted down by adding water at a ratio of 1:1 with dewatering returns • Systems are available which can treat the liquor returns without the use of additional carbon or alkalinity, or pre-treatment or dilution. However, their performance is site-specific
Issues with ammonia toxicity	• Operate downstream anaerobic digestion plant at 35 rather than approximately 40°C to reduce unionized ammonia levels • Run digestion plant at lower retention times (approximately 10 d) to minimize ammonia production and pH increase • Reduce thermal hydrolysis reaction temperature to 150°C to minimize solubilization of proteins • Add un-thermally hydrolyzed high-carbon material to build up intermediate products during digestion to decrease pH. If thermally hydrolyzed, the material is rapidly degraded before a significant pH drop is noticed

REFERENCES

Abelleira J., Pérez-Elvira S. I., Sánchez-Oneto J., Portela J. R. and Nebot E. (2012). Advanced thermal hydrolysis of secondary sewage sludge: a novel process combining thermal hydrolysis and hydrogen peroxide addition. *Resources, Conservation and Recycling*, **59**, 52–57.

Auerbach E. A., Seyfried E. E. and McMahon K. D. (2007). Tetracycline resistance genes in activated sludge wastewater treatment plants. *Water Research*, **41**, 1143–1151.

Banks C. J., Serna Maza A., Zhang W. and Heaven S. (2016). Ammonia Removal in Anaerobic Digestion by Biogas Stripping. Proceedings of SludgeTech Conference, London, UK.

Barber W. P. F. (2016). Thermal hydrolysis for sewage treatment: a critical review. *Water Research*, **104**, 53–71.

Barber W. P. and Stuckey D. C. (1999). The use of the anaerobic baffled reactor (ABR) for wastewater treatment: a review. *Water Research*, **33**(7), 1559–1578.

Barker D. J. and Stuckey D. C. (1999). A review of soluble microbial products (SMP) in wastewater treatment systems. *Water Research*, **33**(14), 3063–3082.

Bergius F. (1932). Chemical reactions under high pressure. In: Nobel Lectures, A. Tiselius (ed.), Chemistry 1922–1941, World Scientific Publishing Company, pp. 245–276.

Bougrier C., Delgenès J. P. and Carrère H. (2008). Effects of thermal treatments on five different waste activated sludge samples solubilisation, physical properties and anaerobic digestion. *Chemical Engineering Journal*, **139**(2), 236–244.

Bouki C., Venieri D. and Diamadopoulos E. (2013). Detection and fate of antibiotic resistant bacteria in wastewater treatment plants: a review. *Ecotoxicology and Environmental Safety*, **91**, 1–9.

Chapman T. and Muller C. (2010). Impact of series digestion on process stability and performance. *Proceedings of the Water Environment Federation*, **2010**(4), 167–178.

Chen R., Yu X., Dong B. and Dai X. (2020). Sludge-to-energy approaches based on pathways that couple pyrolysis with anaerobic digestion (thermal hydrolysis pre/post-treatment): energy efficiency assessment and pyrolysis kinetics analysis. *Energy*, **190**, 116240.

Chynoweth D. P., Srivastra V. J. and Conrad J. R. (1980). Research study to determine the feasibility of producing methane gas from sea kelp. Annual Report for General Electric Company, IGT Project 30502, Institute of Gas Technology, IIT Centre, Chicago, IL.

Cohen A., Breure A. M., van Andel J. G. and van Deursen A. (1980). Influence of phase separation on the anaerobic digestion of glucose, I. Maximum COD-turn-over-rate during continuous operation. *Water Research*, **14**, 1439–1448.

Cohen A., Breure A. M., van Andel J. G. and van Deursen A. (1982). Influence of phase separation on the anaerobic digestion of glucose, II. Stability and kinetic responses to shock loadings. *Water Research*, **16**, 449–455.

Costa J. C., Oliveira J. V., Pereira M. A., Alves M. M. and Abreu A. A. (2015). Biohythane production from marine macroalgae *Sargassum* sp. coupling dark fermentation and anaerobic digestion. *Bioresource Technology*, **190**, 251–256.

Danso-Boateng E., Shama G., Wheatley A. D., Martin S. J. and Holdich R. G. (2015). Hydrothermal carbonisation of sewage sludge: effect of process conditions on product characteristics and methane production. *Bioresource Technology*, **177**, 318–27.

Davies J., (2012) Sanitation: sewage recycles antibiotic resistance. *Nature*, **487**, 302.

Donoso-Bravo A., Pérez-Elvira S., Aymerich E. and Fdz-Polanco F. (2011). Assessment of the influence of thermal pre-treatment time on the macromolecular composition and anaerobic biodegradability of sewage sludge. *Bioresource Technology*, **102**(2), 660–666.

Forster C. F. (1983). Bound water in sewage sludges and its relationship to sludge surfaces and sludge viscosities. *Journal of Chemical Technology and Biotechnology*, **33**(1), 76–84.

Li Y. Y. and Noike T. (1992). Upgrading of anaerobic digestion of waste activated sludge by thermal pretreatment. *Water Science and Technology*, **26**(3–4), 857–866.

Li B., Romero A., Wadhawan T., Tobin M., Manning E., Higgins M.,... and De Clippeleir H. (2019). Recuperative thickening for sludge retention time and throughput management in anaerobic digestion with thermal hydrolysis pretreatment. *Water Environment Research*, **92**, 465–477.

Mabuza N. T., Pocock J. and Loveday B. K. (2005). The use of surface active chemicals in heavy medium viscosity reduction. *Minerals Engineering*, **18**(1), 25–31.

Mallick N. (2002). Biotechnological potential of immobilized algae for wastewater N, P and metal removal: a review. *Biometals*, **15**(4), 377–390.

Neyens E. and Baeyens J. (2003). A review of thermal sludge pre-treatment processes to improve dewaterability. *Journal of Hazardous Materials*, **98**(1), 51–67.

Ngwenya Z., Beightol S., Ngone Oo T., Vega J., Pathak B., Al-Omari A., Zhu K., Wadhawan T., Murthy S. N. and Higgins M. J. (2015). A Stoichiometric A roach to Control Digester Chemistry and Ammonia Inhibition in Anaerobic Digestion with Thermal Hydrolysis Pretreatment: Model Development. Proceedings of WEF Residuals and Biosolids 2015, Water and Environment Federation, Washington DC.

Pei J., Yao H., Wang H., Ren J. and Yu X. (2016). Comparison of ozone and thermal hydrolysis combined with anaerobic digestion for municipal and pharmaceutical waste sludge with tetracycline resistance genes. *Water Research*, **99**, 122–128.

Peng C., Zhai Y., Zhu Y., Xu B., Wang T., Li C. and Zeng G. (2016). Production of char from sewage sludge employing hydrothermal carbonization: char properties, combustion behavior and thermal characteristics. *Fuel*, **176**, 110–118.

Pook M., Mills N., Heitmann M., Panter K. and Walley P. (2013). Exploring the Upper Limits of Thermal Hydrolysis at Chertsey STW. Proceedings of Aqua-Enviro 18th European Biosolids and Organic Residuals Conference and Exhibition, Manchester, UK.

Qiao W., Peng C., Wang W. and Zhang Z. (2011). Biogas production from supernatant of hydrothermally treated municipal sludge by upflow anaerobic sludge blanket reactor. *Bioresource Technology*, **102**(21), 9904–9911.

Qiao W., Yin Z., Wang W., Wang J. and Zhang Z. (2013). Pilot-scale experiment on thermally hydrolyzed sludge liquor anaerobic digestion using a mesophilic expanded granular sludge bed reactor. *Water Science and Technology*, **68**(4), 948–955.

Rus E., Mills N., Shana A., Perrault A., Fountain P., Thorpe R. B., Ouki S. and Nilsen P. J. (2017). The intermediate thermal hydrolysis process: Results from pilot testing and techno-economic assessment. *Water Practice and Technology*, **12**(2), 406–422.

Svensson K., Kjørlaug O., Higgins M. J., Linjordet R. and Horn S. J. (2018). Post-anaerobic digestion thermal hydrolysis of sewage sludge and food waste: effect on methane yields, dewaterability and solids reduction. *Water Research*, **132**, 158–166.

Thompson T. M., Young B. R. and Baroutian S. (2019). Advances in the pretreatment of brown macroalgae for biogas production. *Fuel Processing Technology*, **195**, 106151.

Tong J., Fang P., Zhang J., Wei Y., Su Y. and Zhang Y. (2019). Microbial community evolution and fate of antibiotic resistance genes during sludge treatment in two full-scale anaerobic digestion plants with thermal hydrolysis pretreatment. *Bioresource Technology*, **288**, 121575.

Valo A., Carrere H. and Delgenes J. P. (2004). Thermal, chemical and thermo-chemical pre-treatment of waste activated sludge for anaerobic digestion. *Journal of Chemical Technology & Biotechnology: International Research in Process, Environmental & Clean Technology*, **79**(11), 1197–1203.

Walley P. (2007). Optimising thermal hydrolysis for reliable high digester solids loading and performance. In: 12th European Biosolids and Organic Resources Conference, P. Lowe and N. J. Horan (eds.), 2007, Aqua Enviro, Manchester, UK.

Wang J. (2018). Thermal and Alkaline Hydrolysis of Sewage Sludge and its Resource/Energy Utilization. Presented at a workshop on pre-treatment technologies, at Singapore International Water Week (SIWW), Singapore.

Wang Z., Wang W., Zhang X. and Zhang G. (2009). Digestion of thermally hydrolyzed sewage sludge by anaerobic sequencing batch reactor. *Journal of Hazardous Materials*, **162**(2–3), 799–803.

Wang L., Chang Y. and Li A. (2019). Hydrothermal carbonization for energy-efficient processing of sewage sludge: a review. *Renewable and Sustainable Energy Reviews*, **108**, 423–440.

Ward A. J., Arola K., Brewster E. T., Mehta C. M. and Batstone D. J. (2018). Nutrient recovery from wastewater through pilot scale electrodialysis. *Water Research*, **135**, 57–65.

Weiland P. and Rozzi A. (1991). The start-up, operation and monitoring of high-rate anaerobic treatment systems: discussers report. Water Science and Technology, **24**(8), 257–277.

Wilson C., Murthy S. N. and Novak J. T. (2008). Laboratory-scaled digestibility study of wastewater sludge treated by thermal hydrolysis. *Proceedings of the Water Environment Federation*, **2008**(3), 374–386.

Xue Y., Liu H., Chen S., Dichtl N., Dai X. and Li N. (2015). Effects of thermal hydrolysis on organic matter solubilization and anaerobic digestion of high solid sludge. *Chemical Engineering Journal*, **264**, 174–180.

Yang D., Hu C., Dai L., Liu Z., Dong B. and Dai X. (2019). Post-thermal hydrolysis and centrate recirculation for enhancing anaerobic digestion of sewage sludge. *Waste Management*, **92**, 39–48.

Yuan H., Zhu N. and Song F. (2011). Dewaterability characteristics of sludge conditioned with surfactants pretreatment by electrolysis. *Bioresource Technology*, **102**(3), 2308–2315.

Zhang T., Yang Y. and Pruden A. (2015). Effect of temperature on removal of antibiotic resistance genes by anaerobic digestion of activated sludge revealed by metagenomic approach. *Applied Microbiology and Biotechnology*, **99**(18), 7771–7779.

Index

IWA Publishing's authorised EU representative for General Product Safety Regulations is Diane D'Arras, 15 rue Duret, 75116 Paris, France, e-mail: safety@iwap.co.uk.

Printed and bound by CPI Group (UK) Ltd, Croydon, CR0 4YY
16/06/2026
02140394-0004